I0000984

Francis Henry Champneys

Experimental Researches in Artificial Respiration in Stillborn

Children

and allied subjects

Francis Henry Champneys

Experimental Researches in Artificial Respiration in Stillborn Children
and allied subjects

ISBN/EAN: 9783337182120

Printed in Europe, USA, Canada, Australia, Japan

Cover: Foto ©berggeist007 / pixelio.de

More available books at **www.hansebooks.com**

EXPERIMENTAL RESEARCHES

IN

ARTIFICIAL RESPIRATION

IN

STILLBORN CHILDREN.

EXPERIMENTAL RESEARCHES

IN

ARTIFICIAL RESPIRATION

IN

STILLBORN CHILDREN,

AND ALLIED SUBJECTS.

BY

FRANCIS HENRY CHAMPNEYS,

M.A., M.B. OXON., F.R.C.P.,

OBSTETRIC PHYSICIAN, AND LECTURER ON OBSTETRIC MEDICINE AT ST. GEORGE'S HOS-
PITAL; CONSULTING PHYSICIAN TO THE GENERAL LYING-IN HOSPITAL; EXAMINER
IN OBSTETRIC MEDICINE, MEMBER OF THE BOARD OF THE FACULTY OF
MEDICINE, AND LATE RADCLIFFE TRAVELLING FELLOW IN THE UNI-
VERSITY OF OXFORD; EXAMINER IN OBSTETRIC MEDICINE IN
THE UNIVERSITY OF LONDON.

LONDON

H. K. LEWIS, 136 GOWER STREET, W.C.

1887.

LONDON

PRINTED BY H. K. LEWIS

136, GOWER STREET, W.

TO

JAMES MATTHEWS DUNCAN, M.D., F.R.S.,

WHO BY LABORIOUS RESEARCH IN THE LABORATORY

HAS REMOVED MANY QUESTIONS IN OBSTETRIC MEDICINE FROM THE CHAOS

OF ASSERTION TO THE REGIONS OF EXACTNESS,

These Papers are Dedicated.

PREFACE.

The enquiry contained in the following pages was begun with the object of settling the question debated in the first chapter. The experiments there detailed suggested an extension of the research, which was therefore pursued in several directions.

This work cannot claim the title of "unpractical"—an epithet often applied to work of high, if not immediate and obvious, utility—the closing chapter forbids this claim. On the other hand, immediate practical utility was not the only object of the author, for several facts were elucidated in the course of the enquiry which were of great interest to him at least, but of which he cannot for the present indicate the practical utility.

The facts that his papers have been quoted and criticised on imperfect acquaintance, and that the experiments therein related are periodically repeated on an imperfect scale, without a knowledge of the points already settled by his own enquiries, together with the absence of any English work on the subject, have led him to re: print them in a collected form.

The experiments do not include every question which can be raised on the subject of Artificial Respiration in Stillborn Children, but the author believes them to include all the questions which are capable of settlement by experiments on stillborn children. Other questions are, however, discussed on such evidence as we possess.

60, Great Cumberland Place, London,

May, 1887.

CONTENTS.

CHAPTER I.

THE AMOUNT OF VENTILATION SECURED BY DIFFERENT METHODS OF
ARTIFICIAL RESPIRATION.

PAGE

Introduction—Enumeration of Methods—Classification of Methods
—Description of Methods—Description of Apparatus—Relation
of Experiments—Summary of Results—Remarks—Conclusions   1

CHAPTER II.

THE EXPANSIBILITY OF VARIOUS PARTS OF THE LUNGS.

Introduction—Tables of Results—Remarks—Analysis of Schmitt's
Cases—Analysis of Joerg's Cases—Analysis of Cases of Schmitt
and Joerg—Conclusions—List of Works quoted .   .   . 51

CHAPTER III.

MEDIASTINAL EMPHYSEMA AND PNEUMOTHORAX IN CONNECTION WITH
TRACHEOTOMY.

Table—Remarks—Experiment—Explanation—Analysis of Post-
mortem Records of St. Bartholomew's Hospital and the Hospital
-for Sick Children—Conclusions 68

CHAPTER IV.

MEDIASTINAL EMPHYSEMA AND PNEUMOTHORAX IN CONNECTION WITH
TRACHEOTOMY (continued).

Frequency of Recorded Occurrence since Publication of last Paper
—An Observed Case 81

viii CONTENTS.

CHAPTER V.

CERTAIN MINOR POINTS.

PAGE

Influence of Air in Abdominal Viscera as an Impediment to Artificial
Respiration—Simple Methods recommended to prevent Entrance of Air into the Stomach in Mouth-to-mouth Inflation
tested—Means of securing the Patency of the Upper Air-
passages tested—Description of Signs of Returning Life in a
Case of Still-birth—Conclusions—Works quoted . . . 85

CHAPTER VI.

EXPIRATORY CERVICAL EMPHYSEMA, THAT IS, EMPHYSEMA OF THE NECK
OCCURRING DURING LABOUR AND DURING VIOLENT EXPIRATORY
EFFORTS.

Frequency of its Occurrence—Ætiology—Clinical Course—Pathology
—Theories—Post-mortem Records—Mode of Experiment—Ap-
paratus—Relation of Experiments—Consideration of Experiments
—Conclusions—Works quoted—Addendum 94

CHAPTER VII.

SOME POINTS IN THE PRACTICE OF ARTIFICIAL RESPIRATION IN
CASES OF STILLBIRTH AND OF APPARENT DEATH AFTER
TRACHEOTOMY.

Definition of "Stillbirth"—Various Conditions of Stillbirth—Their
Mutual Relation—Production of the "Necessity of Breathing"
—Asphyxia—Length of Survival of apparently Dead Children—
Prognosis in Stillbirth—Certainty of Death—Perseverance with
Attempts at Resuscitation—Diagnosis of State of Asphyxia—
State of the Pupils—Various kinds of Breathing—Treatment
of Navel-string—Bleeding from Cord—Objects of Artificial
Respiration—Removal of Foreign Bodies from the Air-passages
—Procuring the Patency of the Air-passages—Excitation of the
Circulation—Ventilation of the Lungs—Method of Schultze—
Method of Silvester—Methods of Pacini and Bain—Methods of
Marshall Hall and Howard—Inflation of Lungs—Its Disadvan-
tages—Direct Action of Heat—Summary of Treatment . . 132

EXPERIMENTAL RESEARCHES

IN

ARTIFICIAL RESPIRATION

IN

STILLBORN CHILDREN.

CHAPTER I.

THE AMOUNT OF VENTILATION SECURED BY DIFFERENT METHODS
OF ARTIFICIAL RESPIRATION.

(From Vol. LXIV. of the "Medico-Chirurgical Transactions.")

Introduction.

THE question whether a new-born child shall live or die
is a matter of so great importance that no apology is
needed for such an investigation as is here detailed ; the
answer to this question not uncommonly depends upon
the success or failure of artificial respiration.

In spite of the importance of the subject, nothing is
accurately known with regard to it, and even good text-
books on midwifery pass over it superficially.

It is not proposed to deal at once with the whole matter,
which is far too large for summary treatment ; this object
will probably be better attained by dividing the subject
and treating each division in detail.

B

The present inquiry deals solely with the question,
"What relative amount of air is each of the various
methods capable of introducing into the lungs of a new-
born child which has never breathed?" This question
is as yet unanswered; for the various committees which
have investigated the main question have hitherto confined their attention to adults, and even the interesting
experiments of Behm, which were performed on six
new-born children, dealt with children which had not
breathed in three cases only, in one of which the experiments failed—a point which seems to have affected his
results.

The present investigation, being solely experimental,
elucidates many points in the physiology of the respiration of new-born children.

These experiments, dealing as they do with children in
whom (for instance) the circulation is not proceeding, no
doubt require translation into terms of children in whom
the circulation has not yet ceased.

On the other hand, they eliminate the influence of reflex
action, a most important matter where methods are to be
rigidly tested, and practically of great moment where the
children are in the stage of pale (flabby) asphyxia, in
which reflex action is abolished, and which includes all
cases of real difficulty.

This question of the stage of asphyxia, is a point on
which far too little stress has been laid, the facts being
that almost any form of irritation may be sufficient to
excite inspiratory efforts in a case of the first (livid) stage
of asphyxia, whereas in the second (pale) stage, not only
can reflex action not be relied on, but all forms of irritation are simply useless and waste time.

It is curious to read the correspondence which has
followed the introduction of each new method of artificial
respiration, the writers kindly hastening to contribute

their mite of evidence in favour of the method, in the form of cases absolutely without any trustworthy details of the stage of asphyxia, and therefore absolutely worthless.

It may be perfectly true that a child can be recovered in select cases either by slapping the nates or by Silvester's method, but to put the two methods together is unphilosophical and likely to obscure rather than to elucidate the question. On the other hand, it is certain that in the only important class of cases slapping and all other forms of irritation are simple waste of time.

Many of the accounts with which the papers have been repeatedly inundated, relate to cases which would probably have recovered if simply let alone.

In order to establish a numerical superiority with regard to any one method according to this system, it would only be necessary to apply it to all children born (an argument which applies to the frequent use of the forceps).

The desiderata of a method of artificial respiration are summarised by Behm as (a) ventilation of the lungs, (b) excitation of the circulation, (c) removal of foreign bodies from the air passages. Of these, only the first is here considered.

It has been thought best to give a description of each method from the original source, as even the common methods are most inaccurately known.

It is not, for instance, an edifying spectacle to see one surgeon leaning his whole weight on a patient's abdomen at the same time as another surgeon elevates the arms, in the belief that Silvester's method is being pursued.

No details which secure even a small excess of ventilation of the lungs are unimportant, and a consideration of such details is included in this inquiry.

The number of bodies experimented on is twenty-six,

of which twenty have been used for the investigation of this portion of the subject.

The experiments were begun January 5th, 1878, and ended July 5th, 1880.

All the bodies used were those of children who had never breathed, both in order to keep within the law and also to procure uniform material for investigation.

It will be pointed out that the chest of a child which has breathed differs essentially from one which has never breathed.

It must be observed that a source of error exists in the order followed in the experiment, the subsequent experiments usually succeeding better than the earlier ones, from increasing ventilation of the lungs.

This error has been eliminated as much as possible by varying the order, and by repeating experiments after the lungs have been shown to have become expanded.

Another error exists in the stretching of the pectoral muscles which is liable to occur in Schultze's and Silvester's methods, especially the modifications of the latter.

The following methods have been employed:—

1. Marshall Hall.
2. Howard.
3. Silvester.
4. Pacini.
5. Bain.
6. Schücking.
7. Schüller.
8. Schroeder
9. Schultze.

They may be classified according to the principle on which they depend.

Classification of methods of manipulation according to their mode of action with regard to ventilation of the lungs.

The methods aim at procuring inspiration.

I. Indirectly . . { By elastic recoil of the chest walls { 1. Marshall Hall. 2. Howard.

II. Directly {

A. By traction {

1. Upward traction of the ribs, clavicles, and sternum {
  1. Exerted through the arms; (3) Silvester; (4) Schücking.
  2. Exerted through the shoulders; (5) Pacini; (6) Bain.

2. Upward and outward traction of the lower ribs, and consequent depression of the diaphragm } 7. Schüller.

B. By gravitation and centrifugal force { Elevation of the ribs, clavicles, and sternum, and depression of the diaphragm } 8. Schultze.

C. By flexion { Increasing capacity of cylindrical thorax by curvature of its anterior wall } 9. Schroeder.

Description of methods.

1. *Marshall Hall* (*Lancet*, 1856, March 1, p. 229, "On a New Mode of Effecting Artificial Respiration," by Marshall Hall, M.D., F.R.S., and April 12th, p. 393, "Asphyxia, its Rationale and Treatment"). "Let the patient be placed in the prone position, the head and

neck being preserved in their proper place. The tongue will fall forwards, and leave the entrance into the windpipe free. . . Let the body be now turned gently on the side (through rather more than a quarter of a circle), and the pressure on the thorax and abdomen will be removed, and inspiration . . . will take place ! The expiration and inspiration are augmented by timeously applying and removing alternately pressure on the spine and ribs."

(P. 394.) " Replace the patient on his face, his arms under his head, that the tongue may fall forward and leave the entrance into the windpipe free, and that any fluids may flow out of his mouth ; then 1. Turn the body gradually but completely on the side, and a little more, and then again on the face, alternately (to induce respiration and expiration). 2. When replaced, apply pressure along the back and ribs, and then remove it (to induce further inspiration and expiration), and proceed as before."

2. *Howard* (*Lancet*, 1877, August 11, p. 194). P. 196 : " Seize the patient's wrists, and having secured the utmost possible extension with them crossed behind his head, pin them to the ground with your left hand so as to maintain it. . . . The rest consists in throwing the weight of the body on the lower ribs, and then suddenly relieving the pressure (this of course requires modification in a fœtus). It can be practised before division of the funis or after."

3. *Silvester* ("The True Physiological Method of Restoring Persons apparently Drowned or Dead, and of Resuscitating Stillborn Children," by Henry R. Silvester, B.A., M.D. Lond., 1858). " 1. To adjust the patient's position place the patient on his back with the shoulders raised and supported on a folded article of dress. 2. To maintain a free entrance of air into the windpipe (by drawing the tongue forwards). 3. To imitate the movements of deep respiration raise the patient's arms upwards by the

sides of his head, and then extend them gently and steadily upwards and forwards for a few moments. Next turn down the patient's arms and press them gently and firmly for a few moments against the sides of the chest."

In Fig. 4, p. 17, the operator grasps the arms above the elbows. The arms are not everted.

Also "The Discovery of the Physiological Method of inducing Respiration in Cases of Apparent Death from Drowning, Chloroform, Still-birth, Noxious Gases, &c.," 3rd ed., 1853.

The directions are the same as above, except that (p. 20) the feet are to be secured; the arms are to be kept "stretched steadily" (upwards) for "two seconds" instead of " a few moments." The operator grasps the arms above the elbows, but in the figure he has seized them distally to the elbows (figs. 24 and 25).

4. *Pacini* (" Di un nuovo metodo di praticare la Respirazione artificiale," Firenze, 1867). The feet of the patient being fixed, the operator stands with the head against his own abdomen, and then with his hands takes a firm hold of the upper part of the arms, applying the forefingers behind and close to the armpit, while the thumb is in front of the head of the humerus. Holding the shoulders thus, he pulls them towards him, and then lifts them in a perpendicular direction.

5. *Bain* (*Med. Times and Gazette*, 1868, December 19th, p. 708). 1st method. The fingers are placed over the front of the axillæ, the thumbs over the ends of the clavicles ; the operator then draws the shoulders upwards, and then relaxes his traction.

2nd method. The shoulders are raised by taking hold of the hands and raising the body about a foot off the table, the position of the arms being at about an angle of 45° beyond the head.

6. *Schücking* (*Berl. Klin. Wochenschr.*, 1877, No. 2,

p. 19). The same as Silvester, except that the arms are drawn upwards and outwards.

7. *Schüller* (*Berl. Klin. Wochenschr.*, 1879, June 2nd, p. 319). The operator, standing either at the left side or at the head of the patient, raises the edges of the ribs with his fingers placed beneath them, and then depresses them. The knees should be kept bent to relax the abdominal walls. The manipulation flattens and depresses the diaphragm.

8. *Schroeder* ("Lehrbuch der Geburtshülfe," Bonn, 1874, p. 673) suggests supporting the child by the back only, letting the arms and legs fall backwards (which will produce opisthotonos), and then bending them in the contrary direction (producing emprosthotonos). The latter to produce expiration, the former inspiration.

9. *Schultze* ("Der Scheintod Neugeborener," Jena, 1871, p. 162). The navel string being tied, the child is seized with both hands by the shoulders in such a way that both thumbs lie on the anterior wall of the thorax, both index fingers extend from behind the shoulders into the axillæ, and the other three fingers of both hands lie obliquely along the posterior wall of the thorax. The head is prevented from falling by the support of the ulnar sides of the two hands.

The operator stands with somewhat separated legs, and bends slightly forwards, holding the child as above described at arms' length, hanging perpendicularly (*1st position, inspiratory*).

Without pausing, he swings the child upwards from this hanging position, at arms' length. When the operator's arms have gone slightly beyond the horizontal, they hold the child so delicately that it is not violently hurled over, but sinks slowly forwards and forcibly compresses the abdomen by the weight of its pelvic end (*1st movement, expiratory*).

At this moment the whole weight of the child rests on the operator's thumbs lying on the thorax (*2nd position, expiratory*).

Any compression of the thorax by the hands of the operator must be carefully avoided. The body of the child rests during the first position with the floor of the axilla on the index fingers of the operator exclusively, and no compression should be exercised on the thorax in spite of the support offered by the hands to the head, nor should the thumbs compress the thorax in front.

When the child is swung upwards, the spinal column should not bend in the thoracic but only in the lumbar region, and the thumbs should not at this time strongly press the thorax, but should only support the body as it sinks slowly forward.

The raising of the body as far as the horizontal should be effected by a powerful swing of the arms (of the operator) from the shoulders; but from that point the arms should be raised more and more slowly, and, by means of a delicately-adjusted movement of the elbow-joints and scapulæ on the thorax, the pelvic end of the child should fall gradually over. By this gradual falling over of the child's pelvis over the belly, considerable pressure of the thoracic viscera is exercised both against the diaphragm and the whole thoracic wall. At this point the inspired fluids often pour copiously from the respiratory openings.

After the child has slowly but completely sunk over, the operator again lowers his arms between his separated legs. The child's body is thereby extended with some impetus; the thorax, released from all pressure (the operator's thumbs lying now quite loosely on the anterior wall of the chest), expands by means of its elasticity; but the weight of the body hanging, as it does, on the index fingers of the operator by the upper limbs, and thus fixing the sternal ends of the ribs, is brought into use for the elevation of the ribs with considerable impetus; moreover, the diaphragm descends by virtue of the impulse which is communicated to the abdominal contents. By this means a deep inspiration is quite passively produced (*2nd movement, inspiratory*).

After a pause of a few seconds in 1st inspiratory position, the child is again swung upwards into the previous position (*1st movement, 2nd position, expiratory*), and while it sinks slowly forwards it brings its whole weight to bear on the thumbs, which rest on the anterior thoracic wall, and mechanical expiration again ensues. At this point any inspired fluids always pour copiously from the mouth and nose, and generally meconium from the anus.

The proceeding is repeated eight or ten times a minute, but more slowly when the inspired fluids flow from the mouth and nose.

Description of apparatus.

Tracheotomy was performed, and a cannula tied into the trachea, this cannula being in connection with an india-rubber tube, interrupted by a T-piece closed by a clamp, for the purpose of admitting air when desired. (When this clamp was opened the manometer is said to have been *readjusted*.) The other end of this tube was connected with a V-tube filled with water to a marked point, about half-way up. Inspiration, therefore, produced a rise of the water in the limb of the tube to which the india-rubber tube was attached, and expiration a corresponding fall. The readings in inches refer to the height of the fluid in this limb above the zero or line of original level (the actual height of the column of fluid being double this), and not to the cubic amount of air inspired.

The **V**-tube was not long enough to register more than six or seven inches above the line of zero; when the effect exceeded this, the manometer was readjusted half-way, the subject being held in *statu quo*.

These facts do not vitiate the comparison of the relative inspiratory value of each different method, but they would have to be remembered in calculating the absolute inspiratory value.

Each inch in length of the manometer tube held about 2 c. c.

The results in the same body only are compared.

EXP. 1.—Male child, at eighth month. Craniotomy, January 5th, 1878, 9 p.m.; experiment, January 9th, 10 a.m. (83 hours). Trachea divided, cannula tied in, tube connected with water manometer.

1. Lungs not inflated.

> *a.* Marshall Hall ⎫
> *b.* Silvester   ⎬ $= 0$
> *c.* Schroeder   ⎭
> *d.* Schultze $=$ rise $\frac{1}{4}$ in.

2. Upper and lower parts of trachea connected by a cannula which was tied in, so as to restore the continuity of the trachea. Head bent back, lungs inflated by mouth to mouth method. (N.B.—Considerable force was required.) Experiments failed, the lungs had been burst by mouth to mouth inflation.

EXP. 2.—Male child, prematurely born at seventh month. Partial placenta prævia; heart beating at birth, but no respiratory efforts. Birth at 1 a.m. March 29th, 1878; experiment March 30th, 1 p.m. (36 hours). Tracheotomy, cannula in trachea, connected with mano-meter, lungs inflated through tube.

1. *Marshall Hall.*—Simple change of position in this method produced only the slightest oscillation. Thoracic pressure produced ¼ inch fall. Thoracic and abdominal pressure produced ½ inch fall.

On putting the child into the expiratory position, applying thoracic and abdominal pressure, and connecting the manometer afresh, relaxation of the pressure produced a rise of a quarter of an inch, very slightly increased on turning the child into the inspiratory posture.

2. *Silvester.*—(*a.*) Without previous pressure on thorax or abdomen. Inspiratory movement produced 1½ inch rise. Abdominal and thoracic pressure produced ½ inch fall.

(*b.*) After applying thoracic and abdominal pressure, manometer was readjusted. Inspiratory movement = 2 inches rise.

3. *Schroeder.*—No change except ¾ inch fall with extreme opisthotonos, ¼ inch fall with extreme emprosthotonos.

4. *Schultze.*—(*a.*) Expiratory movement expelled ⅝ inch, inspiratory movement inhaled ½ inch; on hanging the body by the forearms 1⅛ inch was inspired.

(*b.*) Inspiratory movement inhaled 1 inch. On hanging the body by the forearms stretched outwards and backwards, as in Silvester's method, the total inspiration = 4½ inches.

(*c.*) After reinflation of the lungs. Expiration expelled 2 inches. Inspiration by the combined (Schultze-Silvester) method as above (*b*) inhaled 1 inch.

(*d.*) After reinflation of the lungs. Expiration = 2 inches. Inspiration (Schultze-Silvester) = 3½ inches.

After many experiments it was evident that though the maximum and minimum changes varied, the Schultze-Silvester method (as above) produced much more change than other methods.

Maximum inspired.

Order 1. Schultze-Silvester 4½ inches.
 2. Silvester 2 ,,
 3. Schultze 1 ,,
 4. Marshall Hall ¼ ,,
 5. Schroeder fall in all cases.

Exp. 3.—Female child, full time; accidental hæmor-rhage. Born April 6th, 1878, at 10 p.m.; experiment April 8th, 1878, 2 p.m. (40 hours). Tracheotomy, cannula in trachea, lungs inflated through tube of manometer.

1. *Marshall Hall.*—Simple change of posture produced only a slight oscillation. Expiratory posture with thoracic pressure produced ¼ inch fall. Expiratory posture with thoracic and abdominal pressure produced 1 inch fall. Relaxation of this pressure produced a rise of ¼ inch, sometimes very slightly increased, but oftener slightly diminished, by turning child on its back.

2. *Silvester.*—(*a.*) Without thoracic or abdominal pres-sure. Inspiratory movement = 1½ inch rise. Abdominal and thoracic pressure produced ½ inch fall (below zero).

(*b.*) Abdominal and thoracic pressure applied, mano-meter readjusted. Inspiratory movement = 2 inches.

3. *Schroeder.*—Extreme opisthotonos produced ⅛ inch fall. Extreme emprosthotonos produced ⅜ inch fall.

4. *Schultze.*—(*a.*) Expiration = 3 inches fall. Mano-meter readjusted. Inspiration = 1 inch rise. On repeat-ing the experiment several times without readjusting manometer, the column fell 3 inches on expiration, and rose 1 inch on inspiration.

(*b.*) Combined Schultze-Silvester produced 1½ inch rise.

(*c.*) Inspiration (as described by Schultze) produced 1 inch rise. On hanging body by forearms (Schultze-Silvester) the total inspiration reached 1¾ inch.

(*d.*) Extreme opisthotonos in the position of inspiration produced a fall of ½ inch, apparently from overstretching of the anterior abdominal walls, by which they were approximated to the spine.

5. *Howard.*—On relaxing pressure the column rose very slowly ¼ inch, the thoracic walls very slowly altering their shape, their expansion lasting for one minute.

		Maximum inspired.	
Order 1.	Silvester	2 inches.	
2.	Schultze-Silvester	1¾ ,,	
3.	Schultze	1 ,,	
4.	Marshall Hall	¼ ,,	+
5.	Howard ·	¼ ,,	
6.	Schroeder	fall in all cases.	

Exp. 4.—Female child, full time. Born April 14th, 1878, 7 p.m.; experiment April 15th, 3 p.m. (20 hours). Tracheotomy, &c., lungs inflated.

1. *Marshall Hall.*—(*a.*) Simple change of posture produced only slight oscillation.

(*b.*) Expiration with thoracic pressure = 3½ inches fall.

(*c.*) Expiration (subsequently) with thoracic and abdominal pressure = 2 inches fall.

(*d.*) Thoracic and abdominal pressure applied, and manometer readjusted. Relaxation of pressure produced a rise of ¼ inch, which was not increased on turning child into inspiratory posture.

2. *Silvester.*—(*a.*) Without previous thoracic and abdominal pressure only slight oscillation produced.

(*b.*) Thoracic and abdominal pressure applied, and manometer readjusted; result of Silvester's method = only slight oscillation.

3. *Schroeder.*—Extreme opisthotonos produced a slight fall which emprosthotonos increased to 1 inch, and which

diminished to a fall of $\frac{1}{4}$ inch on repeating the opistho-tonos

4. *Schultze.*—(*a.*) Expiratory movement caused $\frac{1}{2}$ inch fall. Manometer readjusted. Inspiratory movement caused $\frac{1}{4}$ inch rise. On repeating this experiment several times the amount of rise and fall increased until inspiratory movement caused 1 inch rise, and expiratory move-ment 2 inches fall.

(*Note.*—The cause of this was probably clearing the air-passages of mucus, and increasing expansibility of the lungs.)

(*b.*) On repeating the above without readjusting the manometer, the column, after many trials, fell 3 inches on expiration, rising only to level on inspiration.

(*Note.*—Thus the total change agrees with that of the first experiment (*a*), amounting to 3 inches in both cases.)

(*c.*) Inspiratory movement combined with hanging child by forearms (Schultze-Silvester) produced a rise of $1\frac{1}{4}$ inch.

(*d.*) Inspiratory movement (ordinary) produced a rise of $\frac{1}{2}$ inch; on hanging child by forearms this was in-creased to nearly 1 inch; on bending the body backwards (opisthotonos) the amount was reduced to $\frac{1}{2}$ inch, apparently from over-stretching of abdominal walls.

Note 1.—The column rose violently with Schultze's method, overshooting the level permanently maintained. This violence was seen in no other method.

Note 2.—During the employment of Schultze's method great care was taken to avoid stretching of the tube; stretching of the tube produced *rise* of the fluid, from rarification of the air.

5. *Howard.*—Inspiration $= \frac{1}{4}$ inch or a little more, the thorax took $\frac{3}{4}$ minute to expand, which it did very slightly.

Maximum inspired.

Order 1. Schultze-Silvester 1¼ inch.
     2. Schultze   1   ,,
     3. Howard   ¼ ,,   +
     4. Marshall Hall   ¼ ,,
     5. Silvester   0
     6. Schroeder   fall in every case.

Exp. 5.—Male child, full time. Placenta prævia; still-born, without inspiratory efforts or artificial attempts. Born May 17th, 1878, 1 p.m. ; experiment May 18th, 4.30 p.m. (27½ hours). Manometer connected. No inflation of lungs.

    1. *Marshall Hall* = 0.

    2. *Silvester.*—Inspiration (1) = 1¼ inch, (2) = 1¾ inch, (3) = 2½ inches. Expiration restored level of manometer.

    3. *Schroeder.*—Inspiration = 0. Expiration = ¼ inch fall.

    4. *Schultze.*—(*a.*) Inspiration varied from 1 to 2½ inches. Expiration restored level of manometer.

    (*b.*) On letting body hang by forearms (Schultze-Silvester), inspiration reached 4 inches.

    (*c.*) Abdominal and thoracic pressure produced not more than ¼ inch fall.

Maximum inspired.

Order 1. Schultze-Silvester 4 inches.
     2. Schultze   ⎰ 2½ ,,
     3. Silvester · ⎱ 2½ ,,
     4. Schroeder   ⎰ 0
     5. Marshall Hall ⎱ 0

Exp. 6.—Male child, full time. Breech presentation. Born June 24th, 1878, 1 a.m.; experiment June 26th, 1 p.m. (60 hours). Manometer adjusted, lungs inflated.

    1. *Marshall Hall.*—Expiration = ¾ inch fall. Inspiration = 0.

2. *Silvester.*—(*Note.*—Very forcibly performed, the force exceeding that which could be safely used to a living child.) Inspiration (1) $= \frac{1}{2}$ inch, (2) $= 1\frac{3}{4}$ inch, (3) $= 2\frac{1}{4}$ inches, (4) $= 3\frac{1}{2}$ inches, (5) $= 4$ inches, (6) $= 4\frac{1}{2}$ inches. Expiration in each case restored the level of the manometer.

3. *Schroeder.*—Opisthotonos produced slight oscillation. Emprosthotonos $= \frac{1}{4}$ inch fall.

4. *Schultze.*—(*a.*) Manometer readjusted during inspiratory position. Expiration $= 1$ inch fall. Inspiration $= \frac{1}{2}$ inch rise.

(*Note.*—The fluid rose to 1 inch, but this was not maintained, the action was rapid and violent in all cases.)

(*b.*) Manometer readjusted while child was held in expiratory position. Inspiration rose gradually from $\frac{1}{2}$ inch to 4 inches, but not more than $1\frac{3}{4}$ inch was maintained. Expiration restored level of fluid.

(*c.*) Manometer readjusted between expiratory and inspiratory movements, so as to equalize pressures. Expiration $= \frac{1}{2}$ inch fall. Inspiration $= 1$ inch rise.

(*d.*) Inspiratory movement completed by hanging body from forearms (Schultze-Silvester). Inspiration $= 2\frac{1}{2}$ inches (permanent). Expiration restored level of fluid.

5. *Howard.*—Produced the merest oscillation.

Maximum inspired.

Order 1. Silvester (forcible) $4\frac{1}{2}$ inches.
     2. Schultze   4   ,,   (level never main-
     3. Schultze-Silvester $2\frac{1}{2}$,,   [tained).
     4. Schroeder
     5. Marshall Hall } 0
     6. Howard

EXP. 7.—Male child, full time (?). Cæsarian section June 24th, 1878, 5.30 p.m.; experiment June 26th, 2.30 p.m. (45 hours).

A. Manometer adjusted ; lungs not inflated.

1. *Marshall Hall.*—Results = 0.

2. *Silvester.*—Inspiration gradually reached ½ inch. Expiration restored level of fluid.

3. *Schroeder.*—Produced slight oscillation.

4. *Schultze.*—The same.

B. Lungs were inflated. No method produced results.

No conclusions can be derived from this series of experiments.

EXP. 8.—Male child at eighth month. Breech presentation. Born October 11th, 1878, 7 p.m. ; experiment October 13th, 11 a.m. (40 hours). Body, which was cold and stiff, thawed in warm water. Manometer adjusted; lungs not inflated.

1. *Marshall Hall.*—Results = 0.

2. *Silvester.*—(*a.*) On raising arms above head, as soon as abdominal and thoracic walls became tight the fluid was depressed. This was repeated several times.

(*b.*) On fixing feet and applying considerable force to the upper limbs, the fluid rose with one bound to 4 inches, falling on expiration to 1 inch (above zero).

(*c.*) On repeating this experiment, the manometer being readjusted, the fluid rose to 5 inches ; expiration restoring its level.

(*d.*) &c., &c. In many subsequent repetitions the fluid rose to 5 inches on inspiration, falling either to 1 inch above zero, to zero, or to 1 inch below zero on expiration.

(*e.*) Strong abdominal pressure produced a further fall of 1 inch.

(*Note.*—Towards the end of this experiment loud whistling was heard at the root of the neck, due to mediastinal emphysema and pneumothorax. This vitiates the experiment.)

3. *Schroeder.*—Opisthotonos and emprosthotonos both produced slight fall of the fluid.

4. *Schultze.*—Inspiration by degrees reached 3 inches. (*Note.*—This is remarkable considering the presence of pneumothorax.)

No results can be deduced from this series of experiments.

Exp. 9.—Male child, full time. Breech presentation, large size. Born November 26th, 1878, 9.30 a.m.; experiment November 27th, 1878, 9.30 a.m. (24 hours). Double talipes varus; spina bifida. Body very cold and stiff, thawed in hot bath. Manometer adjusted, lungs not inflated.

1. *Marshall Hall.*—(*a.*) Body turned on face with block under thorax; thoracic and abdominal pressure produced $\frac{1}{2}$ inch fall; on readjusting manometer and relaxing pressure $\frac{1}{12}$ inch rise; on then turning body into inspiratory position $\frac{1}{4}$ inch rise.

(*b.*) Simple change of posture without pressure produced only slight oscillation.

2. *Silvester.*—A. No pressure on thorax or abdomen. Inspiration (arms rotated *in*wards) $= \frac{1}{4}$ inch. Inspiration (arms rotated *out*wards) $= \frac{1}{2}$ inch. Pressure on thorax and abdomen produced $\frac{1}{4}$ inch fall.

B. Manometer readjusted during pressure on thorax and abdomen. On relaxing this pressure, fluid rose $\frac{1}{4}$ inch. On raising arms rotated *in*wards, slight additional rise was produced. On raising arms rotated *out*wards, the rise reached $\frac{3}{4}$ inch.

c. Legs were held and arms forcibly raised rotated *in*wards, rise = 6 inches. Rotated *out*wards, rise = 8 inches.

(*Note* 1.—The manometer would not register more than 6 inches; the rise of 8 inches was estimated by readjusting manometer midway.

Note 2.—The outward rotation of arms of course rendered the pectoral muscle tenser.)

D. *Marshall Hall.*—Repeated under the more favourable conditions of lungs already expanded, produced no better results than before.

E. *Schultze.*—A. The child being held in inspiratory position. (*a.*) Expiration produced 1 inch fall.

(*b.*) Pressure applied to thorax and abdomen produced a fall of 1 inch; this was not increased by throwing child into expiratory posture.

(*Note.—i.e.* Schultze's expiratory posture produced results equal to pressure on thorax and abdomen.)

B. Child being held in expiratory posture manometer was readjusted. Inspiration rose in subsequent experiments gradually from ½ to 5½ inches. Expiration restored level of fluid.

C. Child suspended by arms at the end of inspiratory movement (Schultze-Silvester). Inspiration = 7 inches.

(*Note.*—Manometer was readjusted at the end of Schultze's inspiratory movement, and before the child was suspended by the arms.)

4. *Howard.*—Inspiration = ⅛ inch rise.

	Maximum inspired.	
Order 1. Silvester (forcible)	8 inches.	
2. Schultze-Silvester	7 ,,	
3. Schultze	5½ ,,	(level never maintained.)
4. Marshall Hall	¼ ,,	
5. Howard	⅛ ,,	

EXP. 10.—Female child, said to have died in utero after a blow on the mother's abdomen two weeks before term. The child was small and seemed more than two weeks before term. Born December 16th, 1878, 5 a.m.; experiment 2.30 p.m. (6½ hours). Manometer adjusted; lungs not inflated.

1. *Marshall Hall.*—(*a.*) Pressure along the back in

the prone position with a block under the thorax produced $\frac{1}{4}$ inch fall.

(*b.*) Manometer readjusted and pressure relaxed; oscillation.

(*c.*) On then turning body into inspiratory posture $= \frac{1}{4}$ inch rise.

(*d.*) Inspiration without previous pressure $= \frac{1}{4}$ inch rise.

2. *Silvester.*—A. No previous pressure on thorax and abdomen.

(*a.*) Inspiration (arms rotated *in*wards) $= \frac{1}{4}$ inch rise.

(*b.*) Inspiration (arms rotated *out*wards) produced at first $\frac{3}{4}$ inch rise, which increased gradually in repeated experiments to $2\frac{1}{4}$ inches.

B. Thoracic and abdominal pressure, manometer readjusted.

(*a.*) Pressure simply relaxed $= \frac{1}{8}$ inch rise.

(*b.*) Inspiration (arms rotated *in*wards) $=$ total rise of $\frac{1}{4}$ inch.

(*c.*) Inspiration (arms rotated *out*wards) $= 2\frac{1}{2}$ inches rise.

(*Note.*—This rise of $2\frac{1}{2}$ inches occurred suddenly, and with a sort of rhonchus as if the air had forced its way past a plug of mucus.)

C. Legs held and arms forcibly elevated.

(*a.*) Arms rotated *in*wards $= 5$ inches rise.

(*b.*) Arms rotated *out*wards $= 5$ inches rise.

D. Abdominal and thoracic pressure; manometer readjusted before each experiment.

(*a.*) Arms rotated *in*wards $= 8$ inches rise.

(*b.*) Arms rotated *out*wards $= 8$ inches rise.

(*Note* 1.—Manometer readjusted midway.

Note 2.—This proves that the external or internal rotation of the arms is not of importance when enough force is applied to raise the shoulders, &c., to their highest possible position.)

3. *Marshall Hall.*—Repeated under the more favourable conditions of inflated lungs. Inspiration $= \frac{1}{4}$ inch rise.

4. *Schultze.*—A. Body held in inspiratory position.

(*a.*) Expiration = oscillation.

(*b.*) Pressure on thorax and abdomen = oscillation.

B. Body held in expiratory position, manometer readjusted. Inspiration gradually increased up to 8 inches. Expiration restored level of fluid.

(*Note.*—At this point a leak occurred, necessitating the cessation of the experiment.)

Maximum inspired.

Order 1. Schultze ⎱ 8 inches (level never main-
 2. Silvester (forcible) ⎬ tained with Schultze's
 ⎰ method).
 3. Marshall Hall. $\frac{1}{4}$ inch.

Exp. 11.—Large male child, full time. Born January 1st, 1879, 10.30 a.m.; experiment January 3rd, 12 noon (49½ hours). Body very cold and stiff, thawed in hot bath. Manometer adjusted; lungs not inflated.

1. *Marshall Hall.*—(*a.*) Thoracic and abdominal pressure in prone position with a block under thorax produced $\frac{1}{4}$ inch fall.

(*b.*) Manometer readjusted, and pressure relaxed, fluid rose $\frac{1}{8}$ inch.

(*c.*) Child turned into inspiratory posture, total rise $= \frac{3}{8}$ inch rise.

(*d.*) Simple change of posture without previous pressure produced $\frac{1}{4}$ inch rise.

2. *Silvester.*—A. No previous pressure—inspiration.

(*a.*) Arms rotated *in*wards = total $\frac{3}{8}$ inch rise.

(*b.*) Arms rotated *out*wards = 1 inch rise.

B. Pressure on thorax and abdomen, manometer readjusted.

(*a.*) Pressure simply relaxed = $\frac{1}{8}$ inch rise.

(*b.*) Arms raised, rotated *inwards* = total $\frac{3}{4}$ inch rise.

(*c.*) Arms raised, rotated *outwards* = total $\frac{1}{8}$ inch rise.

c. Legs held and arms very forcibly elevated, after thoracic and abdominal pressure.

(*a.*) Arms rotated *inwards* = 10 inches rise.

(*b.*) Arms rotated *outwards* = 11 inches rise.

D. Legs held and arms very forcibly elevated without previous thoracic and abdominal pressure.

(*a.*) Arms rotated *inwards* = 9 inches rise.

(*b.*) Arms rotated *outwards* = $10\frac{3}{4}$ inches rise.

(*Note.*—Manometer readjusted midway.)

3. *Marshall Hall.*—Was tried again under the more favourable circumstances of the lungs having become expanded. Pressure on thorax and abdomen, manometer readjusted.

(*a.*) On relaxing pressure, rise = $\frac{1}{4}$ inch.

(*b.*) On turning child into inspiratory posture, total rise = $\frac{7}{8}$ inch.

4. *Schultze.*—A. Child being held in inspiratory position and manometer readjusted between the experiments.

(*a.*) Expiration gradually increased to $3\frac{1}{2}$ inches fall.

(*b.*) Thoracic and abdominal pressure in supine posture produced $2\frac{1}{2}$ inches fall.

B. Child held in expiratory posture, manometer readjusted; child then thrown into inspiratory posture. Inspiration = 7 inches. Expiration restored level of fluid.

c. The same experiment, the child being finally suspended by its arms (Schultze-Silvester). The total inspired was not increased = 7 inches.

Maximum inspired.

Order 1. Silvester (forcible) 11 inches.
2. Schultze-Silvester ⎫ 7 ,,
3. Schultze ⎭
4. Marshall Hall $\frac{7}{8}$,,

Exp. 12.—Full-grown female child. Delivered by forceps, January 17th, 1879; experiment January 24th (7 days). Body, which was cold and stiff, was thawed in hot bath. Manometer adjusted; lungs not inflated.

1. *Marshall Hall.*—(*a.*) Thoracic and abdominal pressure applied in the prone position with a block under the thorax produced a very slight fall.

(*b.*) Manometer readjusted and pressure relaxed, rise of ⅛ inch.

(*c.*) Child turned into inspiratory posture; total rise = a little more than ⅛ inch.

(*d.*) Simple change of posture without previous pressure produced only slight oscillation.

2. *Silvester.*—A. No previous pressure on thorax or abdomen.

(*a.*) Inspiration (arms rotated *in*wards) = ⅝ inch rise.

(*b.*) Inspiration (arms rotated *out*wards) = ⅛ inch rise.

(*Note.*—Something seemed to be obstructing the air passages.)

B. Forcible elevation of the arms, legs being fixed.

(*a.*) Arms rotated *in*wards; the fluid rose on repeated attempts to 7 inches.

(*b.*) Arms rotated *out*wards; on repeated attempts fluid reached 8 inches.

(*Note.*—Manometer readjusted.)

c. Legs not fixed; pressure on thorax and abdomen preceding manipulations.

(*a.*) Pressure simply relaxed = slight oscillation.

(*b.*) Arms rotated inwards = ¾ inch rise.

(*c.*) Arms rotated outwards = 1¼ inch rise.

3. *Marshall Hall.*—Repeated after presumable expansion of lungs.

(*a.*) Pressure relaxed = oscillation.

(*b.*) Child turned into inspiratory posture = ⅛ inch rise.

4. *Schultze.*—A. Child held in inspiratory position. Expiratory movement caused fall, gradually increasing to 1½ inch.

B. Child held in expiratory posture and then thrown into inspiratory position. Inspiration after several attempts reached 7 inches, but not more than 2 inches was maintained; expiration restored level of fluid.

C. The same, the child finally suspended by its arms (Schultze-Silvester). No increase of air *permanently* retained, *i.e.* 2 inches.

Maximum inspired.

Order 1. Silvester (forcible) 8 inches.
    2. Schultze   7 „ (level never main-
                            tained).
    3. Schultze-Silvester 2 „
    4. Marshall Hall   ½ „

EXP. 13.—Large male child. Born February 26th, 1879, 6 a.m.; experiment February 28th, 12.30 p.m. (54½ hours). Body very cold and stiff, thawed in hot bath. *Mercury* manometer, no inflation of lungs.

1. *Marshall Hall.*—Thoracic and abdominal pressure applied in the prone position with a block under the thorax.

(*a.*) Pressure relaxed = slight oscillation.

(*b.*) Child turned into inspiratory posture = no increase of effect.

2. *Silvester.*—A. No previous pressure. Inspiration.

(*a.*) Arms rotated *in*wards = slight oscillation.

(*b.*) Arms rotated *out*wards = slight oscillation.

B. With previous pressure. Nothing but slight oscillation produced.

C. Legs fixed, arms forcibly raised.

(*a.*) Arms rotated *in*wards = ½ inch rise.

(*b.*) Arms rotated *out*wards = ½ inch rise.

3. *Schultze.*—A. Inspiration repeated several times eventually produced a rise of more than 1 inch. Expiration restored level of fluid.

B. The same, the child finally suspended by arms. Inspiration = 1 inch (as before).

4. *Silvester.*—Repeated after some presumable expansion of lungs (legs fixed, arms forcibly raised), produced the same results as above (2 c).

N.B.—Mercury manometer.

Maximum inspired.

Order 1. Schultze ⎫
     2. Schultze-Silvester ⎬ 1 inch.
     3. Silvester (forcible) ½ ,,
     4. Marshall Hall 0

EXP. 14 (G).—Male child, full time, breech presentation. Born November 30th, 1879, 8 a.m.; experiment December 1st, 1.30 p.m. (29½ hours). Body very cold and stiff, thawed before fire.

A. Manometer adjusted, no inflation of lungs. Long wooden pointer stuck into liver, to mark descent of diaphragm.

1. *Silvester* (legs not held). — Inspiration = ¼ inch. Pointer moves slightly upwards, apparently from traction of skin in elevating arms.

2. *Marshall Hall.*—Inspiration = ⅛ inch. No movement of pointer.

3. *Silvester* (forcible, legs held). — Inspiration = 6 inches. No movement of pointer.

4. *Schultze.*—Inspiration = 2 inches. Pointer moves decidedly upwards = descent of diaphragm.

B. To eliminate the action of the expansion of the thorax, a broad band of strapping was applied, completely encircling the chest.

1. *Silvester* (forcible).—Not more than 1 inch could be inspired.

2. *Schultze.*—Results = 0.

c. Strapping removed, lungs inflated; pointer moving strongly upwards.

1. *Silvester* (legs not held).—Inspiration = 3 inches.

2. *Silvester* (forcible, legs held).—Inspiration = 5 inches.

3. *Schultze.*—Inspiration = 2 inches; distinct upward movement of pointer.

D. Strapping reapplied.

1. *Silvester* (forcible).— Inspiration = 1 inch; pointer not moving.

2. *Schultze.*—Inspiration = 1 inch; pointer ascending.

Maximum inspired.

Order 1. Silvester (forcible) 6 inches.

2. Schultze 2 „

These results show that the maximum inspired under Silvester's system was far more affected by the presence or absence of the strapping than under Schultze's system; in other words, that some of the expansion of the lungs in Schultze's system is due to descent of the diaphragm.

EXP. 15 (H).—Male child, placenta prævia. Born February 27th, 1880, 1.30 p.m.; experiment March 2nd, 2.30 p.m. (73 hours). Body very cold and stiff, thawed before fire and accidentally much scorched. Manometer adjusted. Lungs not inflated, but ineffectual attempts at artificial respiration had been made. A pointer was stuck into the liver.

1. *Pacini.*—A. (*a.*) Inspiration = 3 inches.

(*b.*) On raising the shoulders in a perpendicular direction the column sank, showing that the weight of the body of a child does not fix the trunk sufficiently.

(*c.*) On relaxing traction the fluid did not regain the level by ½ inch. '

B. Manometer readjusted.

(*a.*) Inspiration = 3 inches.

(*b.*) On relaxing traction the fluid sank ½ inch below the level, showing an extra ½ inch to have been expired from the lungs.

c, *d*, *e*, *f*, precisely the same.

2. *Bain.*—Repeated experiments produced precisely the same results as the last experiments by Pacini's method.

3. *Schultze.*—Twelve experiments were tried; the amount inspired varied from 1 to 4 inches, the highest level being never maintained, the action being violent. The amount inspired was usually 1½ inch, and expiration usually carried fluid ¾ inch below zero. The actual amount of change of level being thus 2¼ inches. It was observed that when the weight of the body rested solely on the index fingers placed under the arms, ½ inch more was registered than when the other fingers supported the scapulæ. The pointer in the liver showed the diaphragm to descend by Schultze's method only.

4. *Silvester* (feet fixed).—(*a.*) Inspiration = 3 inches. Expiration = ½ inch below zero.

(*b.*) Inspiration = 3 inches. Expiration = ½ inch above zero.

Ten more experiments were tried, the amount inspired gradually diminishing, the pectoral muscles, which had been severely scorched, gradually giving way.

		Maximum inspired.
Order 1.	Schultze	4 inches (level never maintained).
2.	{ Bain { Pacini { Silvester (forcible)	} 3 inches.

EXP. 16 (I).—Male child, stillborn at full time, accidental hæmorrhage. Born May 1st, 1880; experiment May 4th (about 72 hours). Manometer adjusted. Lungs not inflated.

1. *Pacini* (feet fixed).—A. (*a.*) Inspiration = 6 inches.

(*b.*) On raising the arms perpendicularly from the table the column sank (the weight of the body producing little effect).

(*c.*) On relaxing traction the column remained $2\frac{1}{4}$ inches above zero.

B. Manometer readjusted.

(*a.*) Inspiration = 5 inches, $2\frac{1}{4}$ inches having remained in lungs from last experiment.

(*b.*) On relaxing traction the column remained $1\frac{1}{4}$ inch above zero.

C. Manometer readjusted. Three experiments produced identical results with those just described (B).

(*c.*) Inspiration = 5 inches.

(*d.*) On relaxing traction the column remained $\frac{1}{4}$ inch above zero.

Three experiments produced identical results with those just described (*c*, *d*).

(*e.*) Inspiration = 5 inches.

(*f.*) Traction relaxed, level of fluid fell $\frac{1}{4}$ inch below zero.

Three experiments produced an inspiration of 5 inches, the column falling on relaxing traction to zero, or nearly.

2. *Bain.*—1st method produced precisely the same results as the later Pacini experiments (see above), repeated twelve times. 2nd method gave no results whatever.

3. *Schultze.*—Child held in expiratory posture, and manometer readjusted before each experiment.

(a.) Insp. = 2 inches. Exp. = 1 inch below zero.

(b.) ,, = 2 ,, ,, = $\frac{1}{2}$,, ,,

(c.) ,, = 1$\frac{1}{2}$,, ,, = zero.

(d.) ,, = 4 ,, ,, = $\frac{1}{2}$ inch below zero.

(e.) ,, = 4$\frac{1}{2}$,, ,, = zero.

(f.) ,, = 3$\frac{1}{2}$,, ,, = ,,

(g.) ,, = ⎫ ,, = ,,

(h.) ,, = ⎪ ,, = ,,

(i.) ,, = ⎬ gradually increasing to 5 inches ,, = ,,

(j.) ,, = ⎪ ,, = ,,

(k.) ,, = ⎪ ,, = ,,

(l.) ,, = ⎭ ,, = ,,

The highest level on inspiration was never maintained.

4. *Silvester* (feet fixed).—Manometer readjusted after each experiment. (a.) Inspiration = 4 inches. Expiration = $\frac{3}{4}$ inch above zero.

(b.) Inspiration = 4 inches. Expiration = zero.

(c.) Inspiration = 6 inches. Expiration = 1$\frac{1}{2}$ inch above zero.

(d.) Three experiments gave, inspiration = 6 inches. Expiration gradually sank from 1$\frac{1}{2}$ inch above zero to zero.

(e.) Five experiments produced, inspiration = 6 inches. Expiration = zero.

(N.B.—The later experiments would give no suspicion of pneumothorax, which was nevertheless present.)

Maximum inspired.

Order 1. Silvester (forcible) 6 inches.

2. { Pacini / Bain } 5 ,,

3. Schultze 4$\frac{1}{2}$,, (level not maintained).

Exp. 17 (J).—Male child, full time, accidental hæmorrhage. Born May 14th, 11 a.m.; experiment May 15th,

1 p.m. (26 hours). Manometer adjusted. Lungs not inflated.

1. *Pacini.*—(*a.*) Inspiration = 6 inches. Expiration = $\frac{3}{4}$ inch above zero.

(*b.*) Inspiration = 11 inches (manometer readjusted in middle of experiment).

(*c.*) Inspiration = 9 inches. Expiration = 4 inches above zero.

(*d.*) Inspiration = 5 inches. Expiration = 1 inch below zero.

(*e.*) Inspiration = 6 inches. Expiration = zero.

(*f.*) On pressing abdomen column fell 4 inches below zero (*i.e.* total change = 10 inches).

(*g.*) Inspiration = $5\frac{1}{2}$ inches. Expiration = zero.

(*h.*) On pressing abdomen column fell 3 inches below zero (*i.e.* total change = $8\frac{1}{2}$ inches).

(*i.*) Inspiration = 5 inches. Expiration = 2 inches above zero.

(*j.*) Abdominal pressure depresses column to 5 inches below zero (*i e.* total change = 10 inches).

(*Note.*—This method was repeated with the following alteration; the operator stood facing the subject and forced up the shoulders from below. The results were equally good.)

2. *Bain.*—(*a.*) Inspiration = 10 inches. Expiration = zero.

(*b.*) Inspiration = 6 inches. Expiration = 1 inch above zero.

(*c.*) Abdominal pressure depresses column 7 inches (*i.e.* total change 12 inches).

(*d.*) Inspiration = 0 (cannula having become temporarily occluded by pressing against posterior wall of trachea).

(*e.*) Inspiration = 8 inches. Expiration = 2 inches below zero.

(*f.*) Abdominal pressure depresses column 4 inches (*i.e.* total change 14 inches).

(*g.*) Inspiration = $9\frac{1}{2}$ inches. Expiration = $1\frac{1}{2}$ inch below zero.

(*h.*) Abdominal pressure depresses column 5 inches (*i.e.* total change 16 inches).

(*i.*) Inspiration = 8 inches. Expiration = 1 inch below zero.

(*j.*) Abdominal pressure depresses column 5 inches (*i.e.* total change = 14 inches).

(*k.*) Inspiration = 8 inches. Expiration = $1\frac{1}{2}$ inch below zero.

(*l.*) Abdominal pressure depresses column 3 inches (*i.e.* total change = $12\frac{1}{2}$ inches).

3. *Silvester* (feet fixed).—(*a.*) Inspiration = 5 inches. Expiration = $3\frac{1}{2}$ inches above zero. Abdominal pressure depresses column 8 inches (*i.e.* total change = $16\frac{1}{2}$ inches).

(*b.*) Inspiration = 9 inches. Expiration = 1 inch below zero. Abdominal pressure depresses column 6 inches (*i.e.* total change = 16 inches).

(*c.*) Inspiration = 9 inches. Expiration = 1 inch below zero. Abdominal pressure depresses column 5 inches (*i.e.* total change = 15 inches).

(*d.*) Inspiration = 9 inches. Expiration = 1 inch below zero. Abdominal pressure depresses column 5 inches (*i.e.* total change = 15 inches).

4. *Schultze.*

(*a.*) Inspiration = 1 inch. Expiration = zero.

(*b.*)	,,	= 2	,,	,,	=	,,
(*c.*)	,,	= 3	,,	,,	=	,,
(*d.*)	,,	= 4	,,	,,	=	,,
(*e.*)	,,	= 2	,,	,,	=	,,
(*f.*)	,,	= 3	,,	,,	=	,,
(*g.*)	,,	= 3	,,	,,	=	,,

(Highest level never maintained.)

Maximum inspired.

Order 1. Pacini   11 inches.

     2. Bain   10   ,,

     3. Silvester (forcible) 9   ,,

     4. Schultze   4   ,,   (level never maintained).

Exp. 18 (K).—Male child, full time; presentation of hand, foot, and cord in a contracted pelvis; turning, dislocation of vertebræ at root of neck. Born May 27th, 1880, 6 p.m.; experiment May 29th, 10.30 a.m. ($40\frac{1}{2}$ hours). Manometer adjusted. Lungs not inflated.

1. *Bain-Pacini.*—(The methods being practically identical both in manipulation and results, were classed together; the point being that, the feet being fixed, the shoulders were raised directly instead of—as in Silvester's method—indirectly. Manometer was readjusted after each experiment.)

(*a.*) Insp. = 5 inches. Exp. = $\frac{1}{2}$ inch above zero.

(*b.*) ,,   = 5   ,,   ,,   = $\frac{1}{2}$,,

(*c.*) ,,   = 3   ,,   ,,   = $\frac{1}{2}$ below

(*d.*) ,,   = 2   ,,   ,,   = $\frac{1}{2}$ above

(*e.*) ,,   = 2   ,,   ,,   = $\frac{1}{2}$,,

(*f.*) ,,   = 1   ,,   ,,   = $\frac{1}{8}$,,

(*g.*) Abdominal pressure depresses column $1\frac{1}{2}$ inch.

(*h.*) Insp. = 2 inches. Exp. = $\frac{1}{2}$ inch above zero. Abdominal pressure depresses column 1 inch (*i.e.* total change $2\frac{1}{2}$ inches).

The experiments seemed to fail from giving way of the pectoral muscles.

2. *Schüller.*—Elevating the ribs caused *depression* of the column, the ribs could not be elevated without the liver.

3. *Schultze.*

(*a.*) Insp. = 1 inch. Exp. = zero.

(*b.*) ,,   = $1\frac{1}{2}$,, .   ,,   =   ,,

D

(c.) Insp. = $1\frac{1}{2}$ inch.   Exp. = zero.

(d.)   ,,   = 2   ,,   ,,   = $\frac{1}{2}$ inch above zero.

(e.)   ,,   = $1\frac{1}{4}$,,   ,,   = $\frac{1}{2}$,,

(f.)   ,,   = $2\frac{1}{4}$,,   ,,   = $\frac{1}{2}$,,

(*Note.*—Highest level never maintained.)

Repeated experiments produced inspiration gradually reaching $3\frac{1}{4}$ inches, after which the attempts began to fail. Pneumothorax being found at the autopsy.

Maximum inspired.

Order 1. Bain-Pacini 5 inches.

     2. Schultze $2\frac{1}{4}$,,   (not maintained).

     3. Schüller   fall in every case.

EXP. 19 (L).—Female child. Born June 16th, 1880, 6 p.m.; experiment June 17th, 3.30 p.m. ($21\frac{1}{2}$ hours). Manometer adjusted. Lungs not inflated.

1. *Bain-Pacini.*—(a.) Inspiration = 3 inches. Expiration = zero.

(b.) Inspiration = 4 inches. Expiration = $\frac{1}{2}$ inch below zero. Abdominal pressure depressed level $\frac{3}{4}$ inch (*i.e.* total change $5\frac{1}{4}$ inches).

(c.) Inspiration = 4 inches. Expiration = zero. Abdominal pressure depressed level 1 inch (*i.e.* total change = 5 inches). Manometer readjusted.

(d.) Inspiration = 4 inches. Expiration = $\frac{1}{2}$ inch above zero. Abdominal pressure depressed level 1 inch (*i.e.* total change = $4\frac{1}{2}$ inches). Manometer readjusted.

(e.) Inspiration = 3 inches. Expiration = zero. Abdominal pressure depressed level 2 inches (*i.e.* total change = 5 inches). Manometer readjusted.

(f.) Inspiration = 4 inches. Expiration = zero. Abdominal pressure depressed level $\frac{1}{4}$ inch (*i.e.* total change = $4\frac{1}{4}$ inches.

2. *Silvester* (forcible).—Manometer readjusted each time.

(*a.*) Inspiration = 4¼ inches. Expiration = zero. Abdominal pressure depressed level ¼ inch more (*i.e.* total change = 5 inches).

(*b.*) Inspiration = 5 inches. Expiration = zero. Abdominal pressure depressed level ¾ inch more (*i.e.* total change = 5¾ inches).

(*c.*) Inspiration = 5 inches. Expiration = zero. Abdominal pressure depressed level ½ inch more (*i.e.* total change = 5½ inches).

(*d.*) Inspiration = 5 inches. Expiration = zero. Abdominal pressure depressed level ½ inch (*i.e.* total change = 5½ inches).

(*e.*) Inspiration = 5 inches. Expiration = zero. Abdominal pressure depressed level ¼ inch (*i.e.* total change = 5¼ inches).

(*f.*) Inspiration = 5 inches. Expiration = zero. Abdominal pressure depressed level ¼ inch (*i.e.* total change = 5¼ inches).

3. *Schücking.*—Repeated experiments in which Silvester's and Schücking's methods were tried alternately, and also in which the arms were first raised, and then slightly abducted; and thirdly first abducted and then raised simply, showed abduction to be of no additional value.

Thus Schücking's modifications possess no advantage over Silvester's original directions.

With regard to *flexion of the legs*, rather more air was inspired when the legs were straight than when they were flexed.

It is probable that this is due to the fact that where there is no opisthotonos there is no excessive tension on the anterior body walls.

4. *Schultze.*

(*a.*) Insp. = 3 inches. Exp. = zero.

(*b.*) ,, = 4 ,, ,, = ,,

(*c.*) ,, = 3 ,, ,, = ,,

(*d.*) Insp. $= 4$ inches. Exp. $=$ zero.

(*e.*) „ $= 4\frac{1}{2}$ „ „ $=$ „

(*f.*) „ $= 5$ „ „ $=$ „

. ˙(Highest level never maintained).

Maximum inspired.

Order 1. $\left\{\begin{array}{l}\text{Schultze*}\\ \text{Silvester† (forcible)}\\ \text{Schücking}\end{array}\right\}$ 5 inches.

2. Bain-Pacini 4 „

Exp. 20 (M).—Male child, apparently full time; flooding. Born July 5th, 1880, 1 a.m.; experiment July 5th, 2 p.m. (37 hours). Manometer adjusted. Lungs carefully inflated.

1. *Bain-Pacini.*—Manometer readjusted each time. No abdominal pressure used.

(*a.*) Insp. $= 6$ inches. Exp. $= \frac{1}{2}$ inch below zero.

(*b.*) „ $= 6$ „ „ $= \frac{1}{2}$ „

(*c.*) „ $= 6$ „ „ $= \frac{1}{2}$ „

(*d.*) „ $= 7$ „ „ $= \frac{1}{2}$ „

(*e.*) „ $= 5$ „ „ $= \frac{1}{4}$ „

(*f.*) „ $= 5$ „ „ $= \frac{1}{4}$ „

(*g.*) Abdominal pressure depressed level two inches more (*i.e.* total change $7\frac{1}{4}$ inches).

2. *Silvester.*—No abdominal pressure used, feet held.

(*a.*) Insp. $= 6$ inches. Exp. $=$ zero.

(*b.*) „ $= 6$ „ „ $=$ „

(*c.*) „ $= 6$ „ „ $=$ „

(*d.*) „ $= 7$ „ „ $=$ „

(*e.*) „ $= 6$ „ „ $= \frac{1}{4}$ inch below zero.

(*f.*) „ $= 6$ „ „ $= \frac{1}{4}$ „

(*g.*) Abdominal pressure depressed level 1 inch (*i.e.* total change $= 7\frac{1}{4}$ inches).

* Highest level never maintained.

† The highest level of 5 inches was reached once by Schultze's method, five times by Silvester's.

3. *Schücking.*—Six experiments tried ; in all inspiration = 5 inches.   Expiration = zero.

Maximum inspired.

Order 1. $\left\{\begin{array}{l}\text{Bain-Pacini} \\ \text{Silvester (forcible)}\end{array}\right\}$ 7 inches.

2.   Schücking   5   „

Summary of results.

Maximum in inches.

EXPERIMENT 1.

1. Schultze-Silvester ...... $4\frac{1}{2}$
2. Silvester 2
3. Schultze.................. 1
4. Marshall Hall .... $\frac{1}{4}$
5. Schroeder 0

EXPERIMENT 3.

1. Silvester 2
2. Schultze-Silvester ...... $1\frac{3}{4}$
3. Schultze.................. 1
4. Marshall Hall .... $\frac{1}{4}$ (+)
5. Howard.................. $\frac{1}{4}$
6. Schroeder 0

EXPERIMENT 4.

1. Schultze-Silvester ...... $1\frac{1}{4}$
2. Schultze.................. 1
3. Howard .... $\frac{1}{4}$ (+)
4. Marshall Hall .... $\frac{1}{4}$
5. Silvester 0
6. Schroeder 0

EXPERIMENT 5.

1. Schultze-Silvester...... 4
2. Schultze } .... { $2\frac{1}{2}$
3. Silvester } .... { $2\frac{1}{2}$
4. Schroeder } ...... { 0
5. Marshall Hall } ...... { 0

EXPERIMENT 6.

1. Silvester (forcible) ... $4\frac{1}{2}$
2. Schultze.................. $4\frac{1}{2}$
3. Schultze-Silvester...... $2\frac{1}{2}$
4. Schroeder .... }
5. Marshall Hall .... } 0
6. Howard .... }

EXPERIMENT 9.

1. Silvester (forcible) ... 8
2. Schultze-Silvester...... 7
3. Schultze.................. $5\frac{1}{2}$
4. Marshall Hall .... $\frac{1}{4}$
5. Howard .... $\frac{1}{8}$

EXPERIMENT 10.

1. Schultze............ } ... { 8
2. Silvester (forcible) } ... { 8
3. Marshall Hall .... $\frac{1}{4}$

Maximum in inches.

EXPERIMENT 11.

1. Silvester (forcible) ... 11
2. Schultze-Silvester ... 7
3. Schultze 7
4. Marshall Hall......... $\frac{7}{8}$

EXPERIMENT 12.

1. Silvester (forcible)... 8
2. Schultze 7
3. Schultze-Silvester ... 2
4. Marshall Hall......... $\frac{1}{8}$

EXPERIMENT 13 (Mercury).

1. Schultze .... } 1
2. Schultze-Silvester ... } 1
3. Silvester (forcible) ... $\frac{1}{2}$
4. Marshall Hall......... 0

EXPERIMENT 14.

1. Silvester (forcible) ... 6
2. Marshall Hall .... $\frac{1}{8}$
3. Schultze 2

EXPERIMENT 15.

1. Schultze 4
2. Bain................. }
3. Pacini .... } 3
4. Silvester (forcible) ... }

EXPERIMENT 16.

1. Silvester (forcible) ... 6
2. Pacini .... }
3. Bain................. } 5
4. Schultze .... $4\frac{1}{2}$

EXPERIMENT 17.

1. Pacini 11
2. Bain................. 10
3. Silvester (forcible) ... 9
4. Schultze 4

EXPERIMENT 18.

1. Bain-Pacini 5
2. Schultze .... $2\frac{1}{4}$
3. Schüller. 0

EXPERIMENT 19.

1. Schultze .... }
2. Silvester (forcible) ... } 5
3. Schücking .... }
4. Bain-Pacini............ 4

EXPERIMENT 20.

1. Bain-Pacini............ } 7
2. Silvester (forcible) ... }
3. Schücking 5

NOTE.—" Silvester (forcible) " implies the fixation of the feet.

Remarks.

The results of seventeen experiments are recorded, the remaining three experiments having failed.

The summary of results in each case has been determined, not on the principle of average results, but of maximum effect produced.

To either method objections may be urged :

(*a*.) To the system of registering maximum effect, the objection seems to be that such maximum effect may have been produced in one solitary instance. A perusal of the experiments will set this right.

(*b*.) To the system of registering averages, the objection seems to be that the average amount of effect produced does not really represent the series of experiments unless these are more or less uniform, while a casual failure may bring down the average unfairly.

The determination of the value of the various methods of artificial respiration is a task which is eminently unsuited for rigid numerical statistics, even with regard to the small portion of the subject now under consideration.

It is plain from the foregoing experiments that some of the subjects afforded far better results than others; the difference being in some cases due to the development and rigidity of the thoracic walls (which vary greatly even in mature children) ; in others to the presence or absence of mucus in the air-passages, and also no doubt in some cases to the state of preservation. In determining the relative value of the different methods, such subjects as gave high results are plainly much better guides than those which gave low results.

On looking at the table of results, it will be seen that the earlier experiments eliminated three methods as practically useless, viz. those of Schroeder, Howard, and Marshall Hall.

1. *Schroeder.*—This method seems to be based on the *à priori* reasoning that the capacity of a flexible cylinder increases when it is curved, provided that the concave side is not collapsed.

For instance, the capacity of such an instrument as a concertina, to the projecting rings of which a piece of whalebone had been fastened on one side, would be increased by curving it so as to make its outline convex on the side opposite to the whalebone.

In the body of a child, the spinal column would represent the whalebone, and the ribs the rings.

The reasons why the method fails are :

1st. That the anterior body walls become tense and approximated to the spine when the body is bent into a position of opisthotonos, the capacity of the thorax and abdomen becoming diminished.

2nd. That the ribs are not rigid even in children which have breathed.

3rd. But the principal reason lies in the fact that in children which have never breathed, the position of the thorax is one of *expiration,* and not of *inspiration ;* the thoracic walls are completely collapsed, and there is no thoracic cavity or cylinder to deal with. There is nothing to procure descent of the diaphragm, and the thoracic cavity cannot be expanded in any direction by such means.

2. *Howard.*—The above fact of the collapsed state of a child's chest which has never breathed, sufficiently explains the failure of this method. Indeed, for such children certainly, and for all new-born children probably, " Howard's method " (sc. of artificial inspiration) does not exist. For adults the matter is different ; in them the position of thoracic equilibrium is one of moderate inspiration, the chest walls are moreover capable of elastic recoil, and the manœuvre repeats the

mechanism of respiration in birds, in which expiration requires the expenditure of vital force, inspiration being accomplished by the elastic recoil of the chest walls.

The method ought never to have been made to include stillborn children.

3. *Marshall Hall.*—The above remarks apply to this method also, which may be said to have been proved incapable of producing inspiration.

4. *Schüller.*—This method is incapable of producing inspiration; the ribs are hard to separate from the liver, and their want of firmness prejudices any effect which might otherwise be produced; moreover, groping with the finger-tips between the ribs and surface of the liver is a proceeding not altogether without risk. The bent position of the legs was shown (Exp. 19) to be of no additional value, it being probable that in the absence of opisthotonos there is no excessive tension of the anterior abdominal walls which needs avoiding.

We are left, therefore, with Silvester's method (and its modifications by Pacini, Bain, and Schücking) and Schultze's method.

The earlier experiments plainly showed that the weight of a child's body is not sufficient counterpoise to the necessary traction in Silvester's method. In all but the first few cases the feet were fixed, and only in these cases are the results compared with those of other methods. We shall first deal purely with the amount of air inspired by these different methods; reserving our remarks on their mode of action and the small details to be observed in their employment.

It will be seen that in the great majority of cases in which Schultze's method came into competition with Silvester's or its modifications, it produced less effect (Experiments 6, 9, 11, 12, 14, 16, 17, 18); in two cases

Schultze's method produced the greatest result (Experiments 13, 15); in two cases the results of Schultze's method equalled those of Silvester's or its modifications (Experiments 10, 19).

It remains then to compare with each other Silvester's method and its modifications. This was done in five experiments (15, 16, 17, 19, 20).

Pacini's and Bain's methods proved practically identical and will be considered together.

Schücking's method proved practically identical with Silvester's and will be considered together with it.

Silvester's (and Schücking's) methods produced more results than Pacini's and Bain's in two cases (Experiments 16, 19).

Pacini's and Bain's produced more results in one case (Experiment 17).

The results were identical in two cases (Experiments 15, 20).

The practical result is that, provided the feet are fixed and the body properly laid, the mode of seizure (by arms or shoulders) is a matter of no moment.

We now come to the consideration of the mode of action of these methods and of some of the details of their execution.

5. *Schücking.*—This method was tested in Experiment 19, and found to possess no advantage over Silvester's.

6. *Silvester.*—The principle underlying this method and its modifications is the elevation of the ribs, clavicles, and sternum, and consequent enlargement of the cavity of the chest.

It is to be remarked that this group of manipulations, which produces the greatest amount of ventilation, in no way resembles the normal respiration of a child which is almost purely diaphragmatic.

It was shown (by Experiments 9, 10, 11, 12, 13), that,

in the absence of very violent traction, the position of the arms is of much importance; the effect produced when the arm is everted being more than twice as great as when the arm is *inverted*. This is no doubt due to the mode of insertion of the pectoralis major muscle into the outer lip of the bicipital groove; *eversion* naturally renders this more tense. Silvester gives no directions with regard to this point. The arms should certainly be seized above the elbow.

The block beneath the shoulders should only be just so high as to prevent the chin from falling or being bent forwards on the breast; opisthotonos impedes ventilation by tension of the anterior body walls and their consequent approximation to the spine.

7. *Pacini-Bain.* — These methods being essentially identical are considered together. The mode of seizure of the shoulders was shown to be unimportant.

The amount of ventilation produced proved to be, to all practical intent, identical with that produced by Silvester's method.

One small disadvantage seems to be that the operator does not face the subject's face, but views it upside down. This is a matter not directly germane to the present inquiry, but it was thought well to test the question whether an equal amount of ventilation could be procured by reversing the action and elevating the shoulders from below, the operator facing the subject. This was answered affirmatively by Experiment 17, note.

The latter half of Pacini's method and Bain's second method, consisting in raising the arms from the ground and using the weight of the body as a counterpoise, are inapplicable to new-born children, the weight of the body being insufficient to fix it against the traction of the operator (Experiments 15, 16).

In one case the pectoral muscles seemed to give way under the Pacini-Bain manipulations, as if exposed to greater tension than by the Silvester method (Experiment 18).

8. *Schultze.* — In this most ingenious method the shoulder is grasped as in Pacini's method, the ribs, clavicles, and sternum are elevated as in Silvester's method and its modifications (especially Pacini's method), but the counterpoise, or rather the force, is furnished by the action of gravity and so-called centrifugal force, which not only procure elevation of the anterior and upper thoracic walls, but also descent of the diaphragm.

This method, being hard to describe, somewhat complicated, and almost never practised in England, requires special notice.

In the first place the seizure of the shoulders is a most important point, the object being to throw the weight at the end of the inspiratory movement entirely upon the index fingers placed in the axillæ. Attention is directed to this point in Experiment 15, where it is shown that on it depends the entrance of a considerable amount of air. No weight should be supported by the other fingers lying on the back of the thorax.

Again, the violence of its action was noteworthy; the fluid rose violently in the manometer, and never maintained the highest level reached. The following will show the difference between the levels reached and the levels maintained:

Experiment 6 (*a*), 1 inch to $\frac{1}{2}$ inch; (*b*), 4 inches to $1\frac{3}{4}$ inch; Experiment 12 (*b*), 7 inches to 2 inches.

This remarkable fact requires consideration. The rapid jerk with which the fluid fell implies some elastic recoil, either of the chest walls or lung tissue. No recoil of the chest walls was observed, and there are other reasons for thinking that the cause was over-distension of the lung tissue. One reason for thinking this is that it was usually during the Schultze manipulations that the sucking noise was heard at the root of the neck, which was found to have been due to entrance of air beneath the deep cervical fascia into the anterior mediastinum, and sometimes into the pleura. This subject will receive attention hereafter, the only point here insisted on being that this can only be explained by a greater expansion of the chest walls than the lungs were capable of following, and consequent entrance of air into the thoracic cavity by the route of least resistance.

It must be remembered that a small plug of mucus is sufficient to occlude a comparatively important bronchus in a child, and it is suggested that in the presence of a very rapid and powerful blast into the lungs, a neighbouring lobule or bronchus (with its now very delicate walls) may rapidly dilate and press upon the original seat of obstruction, thus completing the occlusion. Under such circumstances the conditions would correspond with those of collapse of part of the lungs with violent inspiratory efforts, which always procure over-distension of other parts of the lungs. Some such explanation seems needed when we remember the incomplete expansion of the lungs proved by dissection. This is a circumstance which, it must be owned, is not in favour of the method.

It must be observed that this may produce the retraction of the abdomen observed by Behm, just as much as occlusion of the glottis by the tongue may produce it.

Schultze's claim to have detected small atelectasies, and to have proved their removal by percussion several days after birth, should be considered in the presence of the actual lungs of a stillborn child (' Der Scheintod Neugeborener,' p. 172).

In the course of the experiments it suggested itself to try suspending the child by its arms instead of by its armpits. Such experiments are designated "*Schultze-Silvester*" in the descriptions. It was found that more air was introduced in this way than by Schultze's method, ·no doubt from greater tension of the pectoral muscles and greater elevation of the shoulders.

The manœuvre is hardly one which could be safely applied to a slippery living child. It is, however, of interest in itself.

The question of the actual descent of the diaphragm was tested by passing a pointer through the abdominal walls into the liver. Descent of the diaphragm naturally implies descent of the liver and ascent of the pointer.

Schultze's method certainly procures descent of the diaphragm, as will be seen from Experiments 14 and 15. No other method does this.

How much ventilation is due to descent of the diaphragm ?

Experiment 14 was designed with a view to answering this question. A broad band of strapping was tightly applied to the thorax so as to completely encircle it, with a view to preventing its expansion ; it was afterwards removed, then reapplied, and the results compared. It is almost needless to say that the strapping seriously hindered the action of Silvester's method, reducing the column of fluid from 6 inches to 1 inch, although much force was applied.

Schultze's method raised a column of 2 inches before the strapping was applied; after the application of the strapping the results became *nil*. The strapping was then removed, and a column of 2 inches raised again. On reapplying the strapping a column of 1 inch was raised, the pointer indicating some descent of the diaphragm.

On blowing directly into the lungs the pointer moved far more than it ever did with Schultze's method. The diaphragm then does descend in Schultze's method, but ventilation depends probably far more on elevation of the ribs, &c., as in other methods.

It is very remarkable how much ventilation is effected by Silvester's method and its modifications, when it is remembered that it is procured by a mode utterly unlike the natural respiration of a new-born infant.

While these experiments were still proceeding, a valuable paper appeared from the pen of Carl Behm ("Die verschiedenen Methoden der kunstlichen Athmung bei Asphyctischen Neugeborenen," 'Zeitschrift für Geburtshülfe und Gynäkologie,' Band v., Heft i., 1880, p. 36). His conclusions are founded on experiments performed on six subjects, only three of whom had never breathed, and in one of the latter the experiments failed. They therefore depend on six cases, only two of which fulfil the necessary conditions.

In some cases his conclusions agree with mine; in some they differ essentially; a difference which I believe to be due to the facts above mentioned. The essential difference between the conditions of equilibrium of a child's chest before it has breathed and after cannot be too strongly insisted upon. This difference is probably established in a very short time after birth, and entirely alters the conditions, especially with regard to the first group of methods, depending on elastic recoil of the chest walls (Marshall Hall, and Howard).

It will be seen that both these methods in my cases amounted to absolute failures, while in Behm's cases they several times surpassed other methods (as those of Schultze, Silvester, and its modifications). If we look, however, at the only two of his cases which concern still-born children who had never breathed (Cases 2 and 6), we find that in Case 2 Marshall Hall's method gave no results (Howard's not being tried); and that in the other case (Case 6) Marshall Hall's and Howard's methods produced hardly any result, standing at the bottom of the list, with the exception of Schüller's method, which is valueless.

Of his other cases the following were the ages at death :—Case 1, one month; Case 3, one day; Case 4, nine days. In two of these cases (1 and 3) the elasticity of the ribs is described as "good," in the other (Case 4) as "weak."

Attention should be especially called to the fact that a child's thorax can be fairly resilient a day after birth. But the question is not what results would be obtained even so short a time as a day after birth, but what amount of ventilation each method will produce in a child which has never breathed. No doubt all methods are far less efficacious in ill-developed and premature, than in well-developed and mature children.

We therefore venture to think that if Behm's cases are properly considered, they tell in favour of the view above enunciated, that all methods depending on the elastic recoil of the chest walls are useless when applied to really stillborn children as means of ventilation of the lungs.

It must always be remembered that a great source of difficulty in the living child (the patency of the air-passages) is secured in these experiments by tracheotomy ; and this point has to be considered in choosing a method or methods for the recovery of a stillborn child, a question which is far from being settled by the present purely experimental inquiry.

Behm recommends (p. 44) the legs to be bent and the feet not fixed (as recommended by Schüller). The bending of the legs I have shown to be of no moment (Experiment 19), the fixation of the legs I have proved to be quite necessary to produce considerable effect.

Behm found, as I did, the Silvester group of manipulations capable of producing greater ventilation than that of Schultze.

It is easily intelligible that the fact of real stillbirth in the subjects experimented on would affect these far less than the methods of Marshall Hall and Howard, and, indeed, be perhaps in their favour.

The methods of faradisation of the phrenic nerves (Hufeland, "Dissertatio vir. electr. in Asphyxia." Göttingen, 1793), and that of Woillez ("Du Spirophore, appareil de sauvetage pour le traitement de l'Asphyxie." Paris, 1876; "Bulletin de l'Acad. de Méd.," Nos. 25, 31, 32, 36, 37, 38; "Comptes Rend.," lxxxii. p. 1447), the principle of which is alternate rarefaction and condensation of the air in the receiver of an air-pump in which the patient is immersed up to the neck, have not been included, the first because it cannot be tried on dead bodies, the second because it requires a large and elaborate apparatus which renders it practically useless.

Conclusions.

The following conclusions are offered :—

1. Since the position of equilibrium of a stillborn child's chest is one of absolute expiration, airlessness, or collapse, no method which depends on elastic recoil of the chest walls will introduce air into its lungs. The methods of Marshall Hall and Howard are useless as means of directly ventilating the lungs of stillborn children.

2. Silvester's method and its modifications by Pacini and Bain introduce more air into the lungs than any other method.

3. In using Silvester's method the arms should be held above the elbows and everted.

4. In using Pacini's or Bain's method the legs should be fixed, the second half of Pacini's method and Bain's

E

second method should not be employed, as the weight of a new-born child's body is insufficient counterpoise to the necessary traction.

5. In using these two latter methods, the operator may face the subject, and lift the shoulders from below; by this means he is able to watch the child's countenance, and is able to introduce an equal quantity of air.

6. Schücking's method is no improvement on Silvester's.

7. Schüller's method is useless and not free from risk.

8. Schroeder's method is useless.

9. Schultze's plan, although its power of ventilation is less than that of Silvester and its modifications, yet acts efficiently.

10. In Schultze's method the diaphragm does descend, though but slightly; its principal action, however, is on the thoracic walls as in the Silvester group.

11. In Schultze's method it is important that the whole weight should rest (at the end of the inspiratory movement) on the index fingers in the axillæ, and should not be distributed to the other fingers.

12. The violence of the action of the method of Schultze is not in its favour.

13. Opisthotonos always produces expiration by tension of the anterior body walls, and should be avoided.

It has been thought unnecessary to include statistics with regard to the expiratory force of each method; considering that direct pressure amounting to any desired force can be applied in all cases where the posture is horizontal. The method of Schultze possesses considerable expiratory force.

CHAPTER II.

THE EXPANSIBILITY OF VARIOUS PARTS OF THE LUNGS.

(From Vol. LXIV. of the "Medico-Chirurgical Transactions.")

Introduction.

THE present communication, being one of a series on the subject of Artificial Respiration in Stillborn Children, proposes to deal with the question, "Which part of the lungs of a child which has been subjected to various methods of manipulation with this object is most frequently expanded and which most frequently unexpanded?"

It would have been interesting to answer this question for each separate method, but material is not plentiful enough. Nothing short of a series of observations on the appearances in a considerable number of cases manipulated by each method would suffice for a satisfactory answer to this subdivision of the question.

Moreover, it is probable that all the methods which are effectual in producing inspiration would be found to produce aëration of very much the same parts of the lungs, since they all act principally in the same way, namely, by raising the ribs, clavicles, and sternum.

Even the method of Schultze, which does produce some descent of the diaphragm, still acts principally in the same way as the Silvester group, as I have elsewhere shown.

It would also be very interesting to determine the order

E 2

in which various parts of the lungs become expanded by inspiring any desired quantity of air.

Unfortunately this is beyond our power; the bodies of stillborn children possess (so to say) individual peculiarities, and some lend themselves much more favourably to experiments than others.

It would have been easy to remove the respiratory organs or open the thorax, and force a measured quantity of air into the chest, observing the order of expansion of the several parts, but the conditions would be so essentially changed, both as regards the method of aëration and the surroundings of the lungs, that the similarity between artifice and nature would cease, and the investigation possess little probable and absolutely no demonstrable value.

This question, however, is more or less elucidated by the following experiments, in which the amount of air changed can be compared with the condition of the lungs as shown by dissection.

In the first table the first column shows the reference to the experiment, the second the methods of manipulation employed, the third the amount of air changed, and the fourth the conditions of the thorax as shown by dissection.

With regard to the amount of air changed, it may be necessary to say that where abdominal and thoracic pressure was employed for the purpose of expelling as much air as possible, the amount of aëration is calculated from the air they expelled as well as from the height of the column of fluid drawn up by inspiration.

This air is of course not under equal pressure, as would have been the case with a spirometer, but the figures are sufficiently near to give an idea of the amount of air received by the lungs in each case.

It may be added that the experiments were instituted

with a view to ascertaining the inspiratory value of the different methods, and have been utilised for the present inquiry.

The use of a spirometer would have prevented several important observations, such as the force with which the air is inspired under Schultze's method.

Although twenty-six subjects were experimented on, only such experiments as refer to our subject are here recorded; and no case has been quoted in which direct inflation of the lungs was practised.

The observations quoted are twelve in number.

The second table, referring to the same experiments, needs no additional explanation.

TABLE I.

No. of Experiment.	Methods employed.	Maximum Inspiratory Effect.	Autopsy.
5	Marshall Hall.........	0	*Left lung.*—All parts contain air; most expanded part, anterior portion of upper lobe; next to this, posterior part of base of lower lobe; least expanded part is the "lingula." *Right lung* more uniformly distended; middle lobe most expanded. Every lobe floats in water. Of the right lung the most buoyant lobe is the middle, then the lower, then the upper; of the left lung the more buoyant lobe is the upper. Several small portions of both lungs do not float alone, among them the left apex. The most expanded portion of the whole lungs is the anterior part of the left upper lobe.
	Silvester .... {	2½ in. 5 cc.	
	Schroeder	0	
	Schultze .... {	2½ in. 5 cc.	
	Schultze-Silvester {	4 in. 8 cc.	
8	Marshall Hall.........	0	Mediastinal emphysema and double pneumothorax. *Left lung* airless. *Right lung.*—All the middle lobe and upper and inner edge of lower lobe fully inflated, the rest airless.
	Silvester (forcible) {	6 in. 12 cc.	
	Schroeder	0	
	Schultze .... {	3 in. 6 cc.	
9	Marshall Hall...... {	¾ in. 1·5 cc.	*Left lung* completely inflated, except a patch along anterior inferior edge, which was partially inflated. *Right lung* completely inflated, but lower lobe less than the rest.
	Silvester (forcible) {	6 in. 12 cc.	
	Schultze .... {	5½ in. 11 cc.	
	Schultze-Silvester {	7 in. 14 cc.	
	Howard .... {	⅛ in. 0·25 cc.	

No. of Experiment.	Methods employed.	Maximum Inspiratory Effect.	Autopsy.
10	Marshall Hall......{	½ in. 1 cc.	*Left lung* floats as a whole, but upper lobe sinks ; lower lobe floats, apex sinks ; lingula floats with difficulty ; lower lobe more buoyant than lingula. *Right lung* floats as a whole ; both lobes (there is no middle lobe) float at first, but upper lobe soon sinks ; left lower lobe is less buoyant than right lower lobe. No lobe of either lung is thoroughly inflated.
	Silvester (forcible){	8 in. 16 cc.	
	Schultze ....{	8 in. 16 cc.	
11	Marshall Hall......{	⅞ in. 1·75 cc.	Mediastinal emphysema, right pneumothorax. *Left lung* floats as a whole ; each lobe floats, the lower lobe the better ; apex is airless ; of lower lobe, lingula floats best, posterior part of base worst. *Right lung* floats as a whole ; upper lobe sinks apex downwards, the other lobes float ; apex airless ; middle lobe floats best ; posterior part of base is airless.
	Silvester (forcible){	11 in. 22 cc.	
	Schultze ....{	7 in. 14 cc.	
12	Marshall Hall......{	⅛ in. 0·25 cc.	Mediastinal emphysema, left pneumothorax. *Left lung* sinks as a whole ; upper lobe airless ; lower lobe floats, but when divided the lingula floats, the rest sinks. *Right lung* floats as a whole, anterior surface uppermost ; all the anterior surface fairly inflated, the rest airless ; lower and middle lobes float ; upper lobe sinks, and is nearly airless.
	Silvester (forcible){	8 in. 16 cc.	
	Schultze ....{	7 in. 14 cc.	
	Schultze-Silvester {	7 in. 14 cc.	
13 Mercury manometer.	Marshall Hall.........{	0	*Left lung* quite airless. *Right lung.*—A few inflated lobules along lower front edge of upper lobe and extreme inner part of this border, and at corresponding point of middle lobe ; lower lobe contains most
	Silvester (forcible){	½ in. 1 cc.	
	Schultze ....{	1 in. 2 cc.	
	Schultze-Silvester {	1 in. 2 cc.	

No. of Experiment.	Methods employed.	Maximum Inspiratory Effect.	Autopsy.
.			air, then upper lobe ; lower lobe partly inflated along anterior surface and along upper and lower edges.
E	(Child laid on its face after death)* Silvester (forcible)...	...	Both lungs mostly inflated, right more than left. *Left lung* all inflated, except a vertical strip lying alongside vertebral column from apex to base, and a small patch of upper lobe just anterior to root. *Right lung* all inflated, except a patch of lower lobe just anterior to root. Each lung floats as a whole, but the above-named airless parts sink when separated.
G (14)	Silvester (forcible) {	6 in. 12 cc.	Mediastinal emphysema, no pneumothorax.
	Marshall Hall.. .... {	⅛ in. 0·25 cc.	Both lungs expanded ; least inflated part being two strips running vertically along the
	Schultze .... {	2 in. 4 cc.	back, opposite the angles of the ribs.
J (17)	Pacini............. {	16 in. 32 cc.	*Left lung* floats ; both lobes separately float ; all parts well expanded except lower and posterior edge of upper lobe and a line along the costal angles of the lower lobe, which is quite airless, and sinks. *Right lung* florid and expanded ; one or two airless patches, the largest along the costal angles of the lower lobe.
	Bain .... {	16 in. 32 cc.	
	Silvester (forcible) {	16 in. 32 cc.	
	Schultze .... {	4 in. 8 cc.	

* The object was to see whether posture had anything to do with the matter, allowing the blood to gravitate to the front instead of to the back of the lungs.

No. of Experiment.	Methods employed.	Maximum Inspiratory Effect.	Autopsy.

K ·(18)	Pacini-Bain ...... {
	Schüller ....
	Schultze .... {
		5 in.
		10 cc.
		0
		2¼ in.
		4·5 cc.
			Mediastinal emphysema, right pneumothorax.
			Left lung.—Anterior inferior part of upper lobe and lingula fairly expanded ; front of lower lobe fairly expanded ; lowest part of base posteriorly fairly expanded.
			Right lung airless.

L (19)	Pacini-Bain ...... {
	Silvester (forcible) {
	Schücking ....
	Schultze .... {
		5½ in.
		11 cc.
		5¾ in.
		11·5 cc.
		Do.
		5 in.
		10 cc.
			Mediastinal emphysema, no pneumothorax.
			Left lung quite airless.
			Right lung airless, except middle lobe and anterior internal part of lower lobe. At anterior internal part of middle lobe is a patch of subpleural emphysema.

TABLE II.

No. of Experiment.	Left Lung.		Right Lung.		
	Upper Lobe.	Lower Lobe.	Upper Lobe.	Middle Lobe.	Lower Lobe.
5	Anterior surface best expanded of all parts of lungs; apex not floating	Posterior part of base well expanded; lingula least expanded	Least expanded	Most expanded	Fairly expanded.
8	Mediastinal emphysema and double pneumothorax.				
	Airless		Airless	Well expanded	Upper and inner edge well expanded; the rest airless.
9	Completely expanded	Completely expanded; except a patch on anterior inferior edge	Completely expanded	Completely expanded	Less well expanded.
10	Sinks in water	More buoyant than upper lobe (lingula hardly floats); less buoyant than right lower lobe	Less buoyant than lower lobe; soon sinks	Absent	More buoyant than right upper or left lower lobe.
11	Mediastinal emphysema; right pneumothorax.				
	Less buoyant than lower lobe; airless	More buoyant than upper lobe; lingula best expanded; posterior part of base least expanded	Sinks in water; apex airless	Best expanded	Floats; posterior part of base airless.

No. of Experiment.	Left Lung.		Right Lung.		
	Upper Lobe.	Lower Lobe.	Upper Lobe.	Middle Lobe.	Lower Lobe.
12	Airless	Sinks as a whole. Floats; but when divided lingula floats, the rest sinks	Mediastinal emphysema; left pneumothorax. Floats as a whole. Nearly airless; part of anterior surface inflated, the rest airless	Floats; anterior surface inflated, the rest airless	Floats; anterior surface inflated, the rest airless.
13	Quite airless		A few inflated lobules along lower anterior edge and inner part	A few inflated lobules at inner extremity of upper anterior border; the least expanded of the three lobes	Partly inflated along anterior surface and upper and lower edges; the best expanded of the three lobes.
E	All expanded except vertical strip along angles of ribs, and a small patch anterior to root	Less expanded than right. All expanded except vertical strip along angles of ribs	Completely expanded	More expanded than left. Completely expanded	All expanded except a patch anterior to root.
G (14)	Mediastinal emphysema; no pneumothorax. Well expanded, except a vertical strip along angles of ribs behind.		Well expanded, except a vertical strip along angles of ribs behind.		

No. of Experiment.	Left Lung.		Right Lung.		
	Upper Lobe.	Lower Lobe.	Upper Lobe.	Middle Lobe.	Lower Lobe.
J (17)	Well expanded, except lower and posterior edge	Well expanded except line along costal angles behind	Well expanded	Well expanded	Well expanded except patch along costal angles.
K (18)	Mediastinal emphysema ; right pneumothorax.				
	Anterior inferior part fairly expanded	Lingula, front and posterior inferior border fairly expanded	Airless	Airless	Airless.
L (19)	Mediastinal emphysema ; no pneumothorax.				
	Airless	Airless	Airless	Partly expanded ; at anterior internal part a patch of sub-pleural emphysema	Airless except anterior internal part.

Mediastinal emphysema occurred in six cases; in four cases with pneumothorax.

Pneumothorax: of the four cases one affected both pleural cavities, two the right, one the left.

In all cases the lung on the side of the pneumothorax was less expanded than the other.

The pneumothorax was probably partly the cause and partly the effect of the complete or incomplete collapse of the lung. Pneumothorax of course impedes the expansion of the lung. On the other hand, the rupture will probably occur into that side on which the lung expands the less, the pressure being greater on that side.

This subject will hereafter be treated more particularly.

In seven cases one lung could be said to be generally better expanded than the other.

In six cases the right was the better expanded, in one the left.

The apices were frequently the least expanded parts of the lungs, *e. g.* No 5 both apices; No. 8 right; No. 10 both; No. 11 both; No. 12 left; L (19) right.

The anterior surfaces were frequently better expanded than the posterior surfaces, *e. g.* No. 5 left upper lobe; No. 12 whole of right lung; No. 13 whole of right lung; K (18) left lung; L (19) middle and lower lobes.

In several of these cases the expanded patch was not bounded by lobes, but extended over the fissures to adjacent lobes, *e. g.* No. 12 whole of right lung; No. 13 right lung; K (18) left lung; L (19) middle and lower lobes.

In several cases an unexpanded strip lay vertically along the costal angles on each side of the vertebral column, *e. g.* E left lung; G (14) both lungs; J. (17) lower lobes both lungs.

In two cases these strips were not limited by lobes but crossed the fissures, *e. g.* E ; G(14).

In one case, J (17), the strip was seen on both lower lobes.

In some cases the lingula was better expanded than the rest of the left lung, *e.g.* 11 ; 12 ; K (18).

In two cases it was airless or little expanded, *e.g.* 5 ; 10.

In no case was the posterior aspect of the lung generally better expanded than the anterior aspect; indeed, in one case (in which, with a view to testing whether the more frequent expansion of the anterior aspect of the lungs was due to supine decubitus after death, and consequent hypostatic gravitation of blood to the posterior aspect), the unexpanded vertical strip along the angles of the ribs was particularly well marked (Exp. E).

On the whole it may be said that the parts of the lungs most frequently and thoroughly expanded were the anterior surfaces, not including the apices or anterior inferior edges.

The parts less frequently and thoroughly expanded were the apices, lower edges, and the vertical strips along the angles of the ribs posteriorly.

The more frequent and thorough expansion of the right lung may be due to the greater directness and size of the right bronchus.

These conclusions are very interesting, and can be accounted for.

First, the fact of the frequent non-expansion of the apices shows that the distance from the trachea does not explain the phenomena. Again, no doubt plugs of mucus will account for some of the unexpanded patches, but they will not explain the greater frequency of their occurrence in certain regions. Moreover, the very distribution of many of these expanded and unexpanded patches disproves such cause, inasmuch as patches due to the patency or closure of a tube would be *limited to a lobule or lobe.*

Now we have seen that these patches are distributed irrespective of fissures and lobes; therefore some other cause must be found.

This cause is, we think, to be found in the mobility of various parts of the chest walls.

The contents of the thorax are not under the conditions of fluid in a vessel, and the laws of equal pressure do not apply. If the chest walls are elevated in one part the external pressure opposite that spot is diminished, and the air will tend to expand the subjacent lung by preference.

On the other hand, the least mobile parts of the chest walls should overlay the least expanded parts of the lungs.

To apply these principles to the matter in hand. The most efficient methods of artificial respiration (viz. the Silvester group and the method of Schultze) act principally by elevating the ribs, clavicles, and sternum.

In a new-born child the upper ribs are hardly at all mobile, and the lower ribs are by far the most mobile. It is beneath the lower ribs, then, that the lungs should be most expanded, and this is the case; the action will be less at the diaphragm than a little remote from it, as the diaphragm acts little in Schultze's methods, and not at all in the Silvester group; and this is also the case. In one instructive case, L (19), the situation selected for a solitary patch of subpleural emphysema was the anterior internal part of the middle lobe of the right lung.

Again the "indifference-point," or centre of motion for each rib is near its angle, and these very points form a vertical line along the spine which we have shown to very usually overlay two unexpanded vertical strips of lung.

How far these observations agree with the appearances in children which have breathed naturally we shall hereafter try to show; meanwhile, it may be remarked that these vertical strips occupy the very same situation as collapsed lung so often does in children, and that the lower edge of the anterior surface of the lungs, occupying as it does the sulcus between the ribs anteriorly and the diaphragm inferiorly, is dependent for its expansion on a consentaneous action of the diaphragm and muscles moving the chest walls, and is often unexpanded. It has been elsewhere remarked that the mode of respiration by any of the efficient methods (Schultze's and the Silvester group), acting as it does by elevating the anterior thoracic walls, differs essentially from the natural mode of respiration in a new-born child, which is all but purely diaphragmatic.

The only considerable collections of the post-mortem appearances of the lungs of new-born children which I have been able to find are those of Joerg and Schmitt.

Out of Schmitt's 101 cases, 36 have been selected as bearing on the question and analysed; out of Joerg's 19 cases, 16 have been selected and similarly treated.

The results are stated in the annexed tables, which have been compiled from positive statements only. The descriptions do not admit of more exact tabulation, being for the most part very brief.

The statements of Paris and Fonblanque, Bednar, Legendre, Gerhardt, Köstlin, Olshausen, Schwartz, and Böhr are either too inexact or are founded on too few cases to carry much weight.

Analysis of Schmitt's cases.

	Best expanded.		Least expanded.
Right lung . . .	11	...	1
Left lung . . .	3	...	4
Anterior surface . .	1	...	0
Posterior surface . .	0	...	2
Right upper lobe . .	9	...	3

	Best expanded.		Least expanded.
Left upper lobe	7	...	4
Right middle lobe	9	...	3
Right base	2	...	12
Left base	3	...	9

Total cases=36 out of 101, no cases in which inflation was practised being included.

Analysis of Joerg's cases.

	Best expanded.		Least expanded.
Right lung	4	...	0
Left lung	1	...	3
Anterior surface	1	...	0
Posterior surface	0	...	1
Right upper lobe	1	...	3
Left upper lobe	2	...	0
Right middle lobe	1	...	3
Right base	1	...	4
Left base	1	...	4

16 cases selected out of 19.

Analysis of cases of Schmitt and Joerg.

	Best expanded.		Least expanded.
Right lung	15	...	1
Left lung	4	...	7
Anterior surfaces	2	...	0
Posterior surfaces	0	...	3
Right upper lobe	10	...	6
Left upper lobe	9	...	4
Right middle lobe	10	...	6
Right base	3	...	16
Left base	4	...	13

52 cases selected out of 120.

The results are that in *children which have died soon after birth :*

F

1. The right lung is usually better expanded than the left.

2. The anterior surfaces are usually better expanded than the posterior.

3. The lower lobes are the parts most often unexpanded.

4. The right middle lobe is usually well expanded.

5. The part of the left lung corresponding to the right middle lobe is often better expanded than the rest of the left lung ; when the right middle lobe, on the other hand, is ill expanded, the corresponding part of the left lung is often unexpanded also (Joerg, No. 1, Case 5 ; No. 2, Case 6).

It is remarkable that in children who have breathed naturally the bases are so often unexpanded; but this seems to correspond with the fact that the characteristic of feeble breathing in infants is feeble descent of the abdominal viscera, betokening feeble descent of the diaphragm.

Conclusions.

The following conclusions are offered :

In stillborn children which have been manipulated by the various efficient methods of artificial respiration :

1. The right lung is more usually and more completely expanded than the left.

2. The anterior surfaces are more usually and more completely expanded than the posterior.

3. The apices of the upper lobes are often unexpanded.

4. The anterior inferior borders the same.

5. One of the places of selection for atelectasis is a strip running vertically along the angles of the ribs on each side of the spine.

6. The patches of expansion or atelectasis, when con-

siderable, are not confined to lobes nor bounded by fissures.

7. The last circumstance is not due to obstruction of the bronchi.

8. The spots of predilection for expansion or atelectasis underlie the regions of greatest and least movement of the chest walls.

List of Works quoted.

Bednar. (1) Die Krankheiten der Neugebornen und Säuglinge. Wien, 1852. (2) Lehrbuch der Kinderkrankheiten. Wien, 1856.

Böhr. Henke's Zeitschr. f. d. Staatsarzneikunde. Erlangen, 1863, Band 85, S. 1.

Gerhardt. (1) Lehrbuch der Kinderkrankheiten. Tübingen, 1861. (2) Id., id., 1878, Band 3, Hälfte 2.

Joerg (Ed.). (1) De morbo pulmonum organico ex respiratione neonatorum imperfecta orto. Lipsiae, 1832. (2) Die Fötuslunge im gebornen Kinde. Grimma, 1835.

Köstlin. Arch. für Phys. Heilkunde. Stuttgart, 1854.

Legendre. Maladies de l'Enfance. Paris, 1846.

Olshausen. Deutsche Klinik., 1864, Band 16, Nos. 36, 37, 38.

Paris and Fonblanque. Med. Jurisprudence, 1823.

Schmitt (Wilh. Jos.). Neue Versuche und Erfahrungen über die Ploucquet'sche und hydrostatische Lungenprobe. Wien, 1806.

Schwartz. Die vorzeitigen Athembewegungen. Leipzig, 1858.

CHAPTER III.

MEDIASTINAL EMPHYSEMA AND PNEUMOTHORAX IN CONNECTION
WITH TRACHEOTOMY.

(From Vol. LXV. of the " Medico-Chirurgical Transactions.")

The occurrence of emphysema in the anterior mediastinum, and pneumothorax, in the course of a series of
experiments on artificial respiration in stillborn children,
has seemed a matter of sufficient interest to justify a short
communication on the subject.

The series of experiments concerned twenty-six subjects
who had never breathed, and extended over a period of
two and a half years, from January, 1878, to July, 1880,
in the laboratory of Dr. Lauder Brunton. They are
related in detail in " Med.-Chir. Trans.," for 1881, pp. 41
–101 (Chapters I. and II. of the present treatise).

Five of the experiments do not concern the present
branch of the inquiry, but in the others the following was
the course pursued:—Tracheotomy was performed, a
cannula was tied into the trachea and connected to an
india-rubber tube, which in its turn was attached to a
water manometer in twenty cases, and to a mercury
manometer in one case. When this was done, the effects
of various methods of respiration by manipulation were
tried, and the height of the column of water noted by the
manometer.

In each case the height noted is the height above zero
or the line of original level, so that the actual height of
the column of water is represented by double the height
indicated.

As the object was that of comparing the relative inspiratory forces of the various methods, this was quite correct as a means of comparison, though the above fact would have to be remembered in calculating the actual inspiratory force.

In the latter case the fact that the manometer would not register more than 6 or 7 inches, and had to be readjusted in the middle of the experiment when the effect exceeded this, would have to be remembered.

The accompanying table shows in the first column the number of the experiment for reference, in the second column the methods of artificial respiration employed, in the third column the maximum inspiratory effect of each method; the fourth column contains the account of the dissection and remarks.

No. of Experiment.	Method employed.	Maximum Respiratory Effect.	Autopsy and Remarks.
8	Marshall Hall ...... Silvester (forcible).. Schroeder .... Schultze.............. No inflation.	0 5 in. 0 3 in.	Loud whistling heard at the root of the neck during the Schultze manipulations. Air in both pleural sacs; thymus embedded in large air-bubbles, which filled the anterior mediastinum, and on pressure escaped from the tracheotomy wound. Left lung airless. Right lung: middle lobe, and upper and inner edge of the lower lobe inflated. No wound in any of the air-passages discoverable.
11	Marshall Hall ...... Silvester (forcible).. Schultze .... Schultze-Silvester... No inflation.	$\frac{3}{8}$ in. 11 in. 7 in. 7 in.	Gurgling and whistling heard during Schultze and Schultze-Silvester manipulations. Air in right pleural sac, none in left. Large bubbles of air in anterior mediastinum, embedding thymus, and extending along course

No. of Experiment.	Method employed.	Maximum Respiratory Effect.	Autopsy and Remarks.
			of left phrenic nerve as far as diaphragm, and along right phrenic vein about half-way to the diaphragm ; on pressure air escapes at tracheotomy wound. Both lungs float, and are generally fairly expanded. No wound in any of the air-passages discoverable.
12	Marshall Hall ...... Silvester (forcible).. Schultze . ... Schultze-Silvester... No inflation.	$\frac{1}{8}$ in. 8 in. 7 in. 7 in.	Whistling heard at root of neck during Schultze manipulations. Air in left pleural sac, none in right. Anterior mediastinum full of air-bubbles, some of which escape on pressure at the tracheotomy wound. Left lung slightly inflated. Right lung better inflated, but only partially.
14 (G)	Silvester (forcible).. Marshall Hall ...... Schultze .... Inflation.	6 in. $\frac{1}{8}$ in. 2 in.	Whistling heard during Schultze manipulations. No air in pleural sacs ; mediastinum full of air-bubbles, embedding thymus. Both lungs fully inflated.
16 (I)	Pacini .... Bain .... Schultze .... Silvester (forcible).. Unsuccessful attempt at inflation.	6 in. 6 in. 5 in. 6 in.	Whistling heard during manipulations (during which is not noted). Air in right pleural sac. Anterior mediastinum full of air-bubbles, which partly escape on pressure into right pleura. Both lungs partly inflated.
18 (K)	Bain-Pacini .... Schüller.............. Schultze .... No inflation.	5 in. 0 $3\frac{1}{4}$ in.	Sucking noise heard during Schultze manipulations, the amount of inspiratory effect diminishing as this increased. Air in right pleural sac, none in left. Anterior mediastinum full of air-bubbles. Right lung airless. Left lung partly inflated.

No. of Experiment.	Method employed.	Maximum Respiratory Effect.	Antopsy and Remarks.
19 (L)	Bain-Pacini .... Silvester (forcible).. Schücking............ Schultze .... No inflation.	4 in. 5 in. 5 in. 5 in.	Sucking noise heard at root of neck during Schultze manipulations. No air in either pleural sac. Anterior mediastinum full of air-bubbles, covering right auricle of heart. Left lung airless. Right lung airless, except middle lobe and anterior internal part of lower lobe. At this point is a patch of sub-pleural emphysema.

Remarks.—It will be seen that mediastinal emphysema was observed in seven out of twenty-one experiments, *i.e.*, in one out of three, or 33·33 per cent.

The first time the phenomena were observed, it occurred for a moment that the emhpysema might be due to post-mortem changes, but the freshness of the subject contra-dicted this.

The next impression was that the air had escaped from one of the bronchi or the trachea, but careful examina-tion and inflation through the trachea with the extre-mities of the bronchi clamped failed to discover any leak.

It was not long before the appearances could be satis-factorily traced to their cause.

Pneumothorax was often associated with mediastinal emphysema, but never occurred without emphysema; on the contrary, the emphysema occurred without pneumo-thorax. Therefore the pneumothorax was probably a later sequel of the emphysema.

Air was observed to escape from the mediastinum into the pleural sac.

Air could be pressed from the anterior mediastinum and emerged at the tracheotomy wound.

A whistling or sucking noise was observed during the experiments at the root of the neck.

It appeared, then, that the air travelled from the tracheotomy wound into the mediastinum, which in some cases it ruptured, producing pneumothorax.

Pneumothorax occurred in five out of seven cases of mediastinal emphysema, in one case into both pleural sacs, in three into the right, in one into the left sac.

In every case but one (16 (1), in which both lungs were about equally inflated), the pneumothorax occurred on the side on which the lung was less expanded.

Experiment.—By way of confirmation a fœtus was taken, and the tissues down to the trachea were divided, the trachea being left intact; the skin was raised so as to form a funnel or pouch, and into this coloured injection was poured.   Silvester's method was then employed; the coloured injection was found in the anterior mediastinum.

The following is the explanation offered :

When by the manipulations for respiration the thoracic cavity is expanded, the intra-thoracic air is rarefied (which is relatively the same thing as the extra-thoracic pressure being increased).

When all the air apertures of the thorax are closed the only way in which the intra- and extra-thoracic pressures can be equalised is by suction through the blood-vessels ; the lungs become hyperæmic, and transudation of serum into the air-passages may occur, causing œdema.

If all the air-passages are not closed the pressure will be equalised through the route of least resistance—the trachea.

This is what occurs in ordinary breathing.

When a manometer is tied into the trachea, the air

enters by the trachea until the column of fluid equalises the pressure, when equilibrium is established.

But the weight of a column of fluid is equivalent to an obstruction of the air-passages; and in this case the same phenomena may be observed in the subject as are observed in a living child which is making violent inspiratory efforts, its air-passages being obstructed, viz. the pressure of the external air depresses the weakest points.

In a living child these include the supra-clavicular fossæ and supra-sternal notch, the abdomen, and hypochondria.

In a dead child the effect of the diaphragm (which is practically not much affected by any artificial mode of respiration) is eliminated, and the hypochondria are not retracted, but the abdomen and parts above the sternum and clavicles are depressed.

But if a potential passage to the thoracic cavity exists, this becomes a *locus minoris resistentiæ*, and the pressure will act through it.

The tracheotomy wound establishes this potential passage by piercing the deep cervical fascia. This fascia, as is well known, runs behind the sternum and becomes continuous with the sheaths of the great vessels and the pericardium.

It is through this passage that air penetrates into the mediastinum.

Whether it goes further than this is determined by the question whether a sufficient quantity can be contained there to equalise the pressure ; if not it will go farther.

In one case it followed the course of the phrenic nerves, on one side to the diaphragm.

In the majority of cases it burst the mediastinum and distended the pleural sac.

It was observed that in most cases one lung was better

expanded than the other, and it was on the less expanded side that the pneumothorax occurred.

This is what we should have expected from the above-named physical conditions.

No case occurred in which it burst into the better expanded side, though in one case, the lungs being (as far as could be seen) about equally expanded, pneumothorax occurred on one side only. (A thickened pleura, however, might determine rupture into the opposite or better expanded side.)

In one case (19 L) it was noted that with mediastinal emphysema there was one patch of subpleural emphysema on the solitary expanded portion of either lung.

With emphysema in connection with partial collapse, solidification, &c., we are familiar; it is but an illustration of the same physical laws as preside over mediastinal emphysema.

It was observed that the noise at the root of the neck in six cases out of the seven was found to occur during the Schultze manipulations; in the seventh case it was heard, but its exact time of occurrence is not noted.

This coincides with what we know of the suddenness and violence of the action of this method noted elsewhere; it was the sudden jerk caused by this method which produced the sudden blast at the tracheotomy wound which burrowed into the anterior mediastinum.

It is to be remarked that the presence of a large mediastinal emphysema, even with pneumothorax, does not necessarily prevent considerable inspiratory effect.

I have looked through the post-mortem records of St. Bartholomew's Hospital from 1868, and find records of twenty-seven cases of autopsies after tracheotomy for various causes.

Mediastinal emphysema is recorded in three cases, or 11·1 per cent.

In two out of these three cases pneumothorax is noted, and in one it was the cause of death. In one case in which pneumothorax was probably bilateral, both lungs were collapsed ; in another, the right-sided pneumothorax corresponded with right-sided collapse.

In one case there was emphysema of the tissues about the neck. In one case the air channel was traced down from the wound into the anterior mediastinum.

I have also examined the post-mortem records of the Hospital for Sick Children since 1860, and find records of eighty-two cases of autopsies after tracheotomy for various causes.

Mediastinal emphysema is recorded in five cases, or 6·09 per cent.

In no case is pneumothorax noted.

The state of the lungs was as follows :

Case 1. Both lungs collapsed.

2. No collapse.

3. Both lungs partially hepatised.

4. Anterior parts of lungs emphysematous, posterior parts less expanded than anterior.

5. Upper lobe and posterior part of lower lobe of right lung solid and airless.

Emphysema of the neck is noted in three out of the five cases.

I do not imagine that these numbers represent the actual facts; indeed, in the cases in which mediastinal emphysema is recorded, it is almost invariably mentioned by the way as if an unimportant fact; and one recent case was only recovered from oblivion by personal inquiries.

This is not to be wondered at, considering that, as far as I have been able to discover, the matter is not even hinted at in any of the books dealing with tracheotomy; indeed, the following passage is the only mention of the subject which I have been able to find :

Wilks and Moxon, "Lectures on Pathological Anatomy," second edition, 1875, p. 308 :

"We believe we have seen two cases of pneumothorax arise from tracheotomy, and we mention the circumstance because we are not aware that it has ever been alluded to. In one case where, after tracheotomy, death occurred without sufficient reason, both lungs were found contracted in the chest, and the cellular tissue in the posterior mediastinum was filled with air, producing large bubbles, which we think had burst through the pleura into the chest.

" In another case, where most extensive superficial emphysema followed the operation, the breathing became laborious before death, and the lungs were found contracted in the same manner ; the emphysema had penetrated the mediastinum."

The latter of these two cases, together with my experiments, may illustrate the production of emphysema of the tissues of the neck, &c., during labour.

With regard to pneumothorax, any small amount of air in the pleura is liable to escape observation unless the thorax is opened under water.

Still, thus much is proved :

1. That mediastinal emphysema does occur after tracheotomy.

2. That it is sometimes accompanied by pneumothorax.

3. That it sometimes exists apart from emphysema of the neck.

The route which the air takes was proved by experiments to be behind the deep cervical fascia.

It remains to discuss the relation of mediastinal emphysema to emphysema of the neck.

It would appear, *à priori*, probable that these would be due to a common cause.

The experiment in which coloured injection penetrated

beneath the deep cervical fascia to the anterior mediastinum, without tracheotomy; and all the experiments
in the table (in which no escape of air from the trachea
was possible) prove that the air does not penetrate from
the tracheal wound.

But if the above views are correct, the two sorts of
emphysema are obviously due, not only to different, but
to opposite causes, mediastinal emphysema being due to
the diminished pressure in the thorax during inspiration,
emphysema of the neck, on the contrary, being produced
during expiration.

Emphysema of the neck may, indeed, run beneath the
skin and fascia upwards to any extent, and downwards
over the thorax, but the air cannot be forced by the
expiratory force into the thorax, for the simple reason
that the *vis a tergo* of the air issuing from the tracheal
wound is at once met by the equally powerful *vis a fronte*
of the intra-thoracic pressure.

The production of anterior mediastinal emphysema and
the route taken by the air need not be again explained.

It is not to be denied that air blown from the tracheal
wound beneath the cervical fascia may be subsequently
sucked into the anterior mediastinum (though air in the
subcutaneous cellular tissue could hardly do so), but what
is insisted on is that, whether the air travels directly
from the wound or indirectly, as explained above, the
mechanism of the production of the two sorts of emphysema is entirely different, and, indeed, opposite.

The following case shows that blood or, indeed, any
other fluid, effused beneath the deep cervical fascia, may
subsequently be drawn into the thorax, a fact which,
perhaps, deserves attention with regard to the after
treatment of operations near the root of the neck and of
deep cervical abscesses. In view of the above observations, strong or sudden inspiratory efforts (*e.g.* in vomit-

ing) must greatly increase the risk of thoracic complications from the inspiration of pus or other fluids into the thoracic cavity.

The case is briefly as follows :—Induction of premature labour for flat pelvis; turning; delay of after-coming head at brim ; child stillborn, with heart beating ; artificial respiration by direct inflation and the methods of Silvester and Schultze for three hours, after which child breathed naturally ; death six hours after birth.

Autopsy showed separation of fifth and sixth cervical vertebræ without fracture, blood extravasated in all directions, among others into the tissues of the neck, including deep cervical fascia, from whence it extended downwards to the apices of the lungs and along the sheath of the vessels on the left side to the level of the third dorsal vertebra. Bloody fluid in both pleuræ.

It is of some consequence to consider the exact moment at which the danger of mediastinal emphysema occurs.

The conditions favouring its production are two—a wound of the deep cervical fascia, and obstruction to the air-passages; the moment of its production is inspiration.

To apply this : the danger begins as soon as the operator has divided the deep cervical fascia. The lower this is divided the more easily emphysema of the anterior mediastinum is produced.

Supposing obstruction to exist in the larynx, the period of greatest danger is the interval between the division of the deep cervical fascia and the establishment of efficient patency of the trachea.

Elevation of this fascia away from the trachea renders the entry of air beneath it more easy.

Bungling in inserting the tube increases the danger by prolonging the dangerous period.

Insertion of the tube beneath the deep cervical fascia instead of into the trachea (of which instances are on record) renders the entrance of air easy above all other conditions.

The following practical conclusions are offered : —

1. Emphysema of the anterior mediastinum occurs in a certain number of tracheotomies.

2. It is often associated with pneumothorax, to which it stands in causal relation, and which may be the cause of death after tracheotomy.

3. The air is most likely to burst into that pleura of which the lung is the less expanded. On the other hand, pneumothorax, of course, helps to collapse the lung.

4. The route selected by the air is the space beneath the deep cervical fascia.

5. Emphysema of the anterior mediastinum may or may not be associated with emphysema of the neck; but their causes are different, and the conditions of their production are opposite.

6. The conditions favouring the production of mediastinal emphysema are division of the deep cervical fascia, obstruction to the air-passages, and inspiratory efforts.

7. The dangerous period during tracheotomy is the interval between the division of the deep cervical fascia and the efficient introduction of the tube.

8. The incision in the deep cervical fascia should not be longer than necessary in the direction of the sternum. It should on no account be raised from the trachea, and this should be particularly remembered during inspiratory efforts.

9. It will probably be found that the frequency of occurrence of emphysema of the anterior mediastinum depends much on the skill of the operator, especially in inserting the tube.

10. If artificial respiration should prove necessary, the tissues should be kept in apposition with the trachea,

and any manipulations performed steadily and without jerks.

11. Schultze's method (which is not otherwise suitable for the above purpose) is especially prone to produce emphysema of the anterior mediastinum.

12. These observations illustrate the fact that, apart from the question of tracheotomy, the inspiratory force of the thorax should be remembered in all operations near the root of the neck, whether dealing with vessels or not, and in the case of all collections of pus beneath the deep cervical fascia. In these cases quiet respiration is essential for the safety of the patient, and vomiting, which begins with a sudden inspiration, is dangerous.

13. These observations may serve to illustrate the production of emphysema of the neck, &c., during labour.

CHAPTER IV.

MEDIASTINAL EMPHYSEMA AND PNEUMOTHORAX IN CONNECTION
WITH TRACHEOTOMY (*continued*).

(From Vol. LXVII. of the " Medico-Chirurgical Transactions.")

IN vol. lxv. of the " Medico-Chirurgical Transactions,"
1882, p. 81 (p. 75 of the present treatise), I wrote as
follows :

" I have examined the post-mortem records of the
Hospital for Sick Children since 1860, and find records
of eighty-two cases of autopsies after tracheotomy for
various causes.

" Mediastinal emphysema is recorded in five cases, or
6·09 per cent. In no case is pneumothorax noted. . . .
I do not imagine that these numbers represent the actual
facts ; indeed, in the cases in which mediastinal emphysema is recorded, it is almost invariably mentioned by
the way as if an unimportant fact ; and one recent case
was only recovered from oblivion by personal inquiries.

" This is not to be wondered at, considering that, as
far as I have been able to discover, the matter is not even
hinted at in any of the books dealing with tracheotomy ;
indeed, the following passage is the only mention of the
subject which I have been able to find." (Then follows
a quotation from Wilks and Moxon, " Lectures on Pathological Anatomy," second edition, 1875, p. 308.)

On November 15th, 1883, I received the following

statement from Dr. Angel Money, Registrar of the Hospital for Sick Children :

" Since the publication of the above paper there have been 28 cases in which an autopsy was made after tracheotomy ; of these 14 were males, 14 females.

"In all cases the examination was made under water.

" In 16 cases out of the 28 (8 males and 8 females), emphysema of the mediastinum was found.

" In 2 of these cases pneumothorax was also found. It was found in no case without emphysema of the mediastinum. The amount of emphysema of the mediastinum was greatest when pneumothorax existed also.

" In many, if not all, cases artificial respiration had been performed."

It will be seen that the occurrence of emphysema was noted in 5 cases out of 82, or 6 per cent. of those cases which ended fatally after tracheotomy in twenty-one years before the publication of the paper above referred to ; and in 16 cases out of 28, or 57 per cent. in the two years following its publication.

Pneumothorax was not noted in a single one of the 82 cases occurring in the twenty-one years previous to the paper, but has been noted twice in 28 cases occurring in the two years following its publication.

These facts speak for themselves.

It seems to be evident that emphysema of the anterior mediastinum is a frequent occurrence in fatal cases of tracheotomy.

What its frequency is in cases which do not end fatally cannot be deduced from the above facts ; but the. occurrence of mediastinal emphysema can hardly be a matter of indifference, and pneumothorax must be a very serious complication.

It is possible that the bubbles of air in the mediastinum may produce pneumothorax by being ruptured into the pleura by pressure on the sternum and thorax. Such pressure usually (and quite rightly) forms part of the means of recovery where artificial respiration is performed.

It is strange that the occurrence of both complications should have been (but for the casual allusion referred to) completely overlooked until the experiments which I have had the honour of communicating to the Society.

In the discussion on the present communication (see "Proceedings of Medico-Chirurgical Society," new series, Vol. I., No. 4, p. 182) Dr. Douglas Powell said that in seven cases of diphtheria dying after tracheotomy, which he had observed in the last two years, two cases of emphysema had occurred. In one of the cases the tube had not been put into the trachea, but along its side. In the other case, dying some hours after the operation, emphysema with collapse of the lung was found. In this case the trachea was opened one inch below the cricoid cartilage, general emphysema ensued, and on auscultation over the mediastinum, the air could be heard penetrating with a dry crepitating sound. In these cases the conditions of Dr. Champneys were very accurately observed, there being (1) strong inspiratory efforts, (2) a difficulty in the entry of air to the lungs, (3) tracheotomy with opening of the deep fascia at the root of the neck. In living children, moreover (and in healthy adults), there was a constant aspiration towards the chest cavity by virtue of the residual elastic tension of the lungs ; which aspiration did not exist in the stillborn subjects of Dr. Champneys' experiments. An incision into the deep tissues near the episternal notch during life (and perhaps even after death), must be attended with entry of air into the mediastinum, and its further accumulation must continue even after opening the trachea, for whilst the indraught to the mediastinum was unopposed, that to the lungs was opposed by their elasticity, which was set free by the entry of air into the thoracic cavity. These considerations had two important bearings on surgical practice—first, it suggested the superiority of the high operation because the cervical fascia was less cut into; and, secondly, the danger, in forcibly throwing back the head in order to make the trachea more prominent, of placing the fascia on the stretch, and of cutting into it within the sphere of lung traction.

CHAPTER V.

CERTAIN MINOR POINTS.

(From Vol. LXVII. of the " Medico-Chirurgical Transactions.")

THE following communication concerns certain minor details in the mode of performing artificial respiration in stillborn children, which it has been thought well to test by experiment.

i. It was decided to test the effect produced by the presence of air in the abdominal viscera on the amount of air capable of being drawn into the lungs by the manipulative methods.

Experiment 1.—In a female child ("Med.-Chir. Trans.," vol. lxiv. 1881, p. 71 ; and in the present treatise, p. 34, Exp. 19 (L)), in which the manipulative methods (Silvester, Pacini, Bain, Schultze) succeeded in effecting considerable inspiration, a cannula was passed through the abdominal walls at the navel, and tied in ; attached to the cannula was an india-rubber tube with a clip, so that the abdominal cavity could be inflated and the air let out.

The quantity of air inspired by these methods was noted.

It was found that moderate inflation of the abdominal cavity produced no effect on the amount of air changed by Silvester's method, and its modifications by Pacini

and Bain. When the abdomen was very tensely inflated, the amount of air inspired was very slightly
diminished.

Experiment 2.—In a male child (see experiments above
quoted, No. 20 (M), p. 73; and in the present treatise
p. 36), the same method of experiment was adopted.

(*a*) The *Pacini-Bain* modification of Silvester's method
was repeated six times.

In each case 6 inches (*i.e.* 6 inches on the water manometer) was indicated as the inspiratory value.

The abdomen was then inflated as tightly as possible,
and the same method repeated six times.

In each case 6 inches was indicated as the inspiratory
value.

(*b*) *Schultze's* method gave 7 inches as the inspiratory
value on six repetitions.

The abdomen was then inflated as tightly as possible,
and the same method repeated six times.

In each case 7 inches was indicated as the inspiratory
value.

Remarks.—The question here was not whether air in
the stomach and abdominal viscera impedes respiration
when naturally conducted, but whether it impedes *artificial respiration*. These experiments answer the question
decidedly in the negative.

The very slight difference noticed in Experiment 1 was
probably due to the impeded movements of the thoracic
walls, produced by great distension of the abdominal
walls.

It must of course be remembered (loc. cit., vol. lxiv.
p. 79; and in the present treatise p. 42), that : " This
group of manipulations (Silvester's and its modifications),
which produces the greatest amount of ventilation, in no
way resembles the normal respiration of a child, which
is almost purely diaphragmatic ;" and that the presence

of air in the abdominal viscera would certainly impede the descent of the diaphragm in a child breathing normally. This, however, is not the question, which really concerns the resuscitation of a stillborn child by artificial respiration.

It must also be remembered that nothing is easier than to empty such a child's stomach of air by gentle pressure.

Conclusion.—The presence of air in the abdomen in no way impedes the ventilation of the lungs by artificial respiration.

ii. It was next determined to test simple methods usually recommended to prevent the entrance of air into the stomach in mouth-to-mouth inflation.

A. It has been recommended to press the larynx, and especially the cricoid cartilage, against the vertebræ, with the idea that the complete cartilaginous ring of the cricoid cartilage will be strong enough to compress the œsophagus without itself collapsing.

This method has been recommended by *Paul Scheel* in 1800, *Herholdt* in 1803, *Olshausen* in 1864, and also by the Committee of the Royal Medical and Chirurgical Society in 1862, who declared that it was competent to effect its object; this, however, applied to adults.

B. It has also been recommended to bend the head strongly backwards, apparently with the idea that the soft œsophagus, lying posteriorly, might become bent in such a way as to be impervious, while the trachea provided with its rings would remain patent.

These points were tested in two experiments.

Experiment 3.—A tube was connected with the pylorus and placed under water, so that bubbles might announce the repletion of the stomach with air. The trachea was connected with a manometer.

A. Mouth-to-mouth inflation was then tried, pressure being put on the cricoid. This was repeated many times.

It was found only occasionally possible to put pressure on the cricoid so as to occlude the œsophagus, without also occluding the trachea.

· B. Bending the head back was found to produce no effect in preventing the entrance of air into the œsophagus.

Experiment 4.—A similar method was adopted.

A. Pressure on the cricoid cartilage occluded either œsophagus and trachea or neither. No graduation of the pressure on the cricoid was able to effect a closure of the œsophagus alone.

B. Bending the head back produced no effect.

Conclusions.—Pressure on the cricoid and bending the head back are powerless to prevent air from passing into the stomach if the lungs are inflated.

iii. The next experiment concerned the patency of the upper part of the air-passages.

Howard had already shown that the common measure of pulling the tongue forward with forceps is useless, as it does not affect the structures at the base of the tongue, and does not raise the epiglottis.

His manœuvre of tilting up the chin with the mouth closed, however useful in adults, permitting respiration solely through the nose as it does, cannot be of the same use in stillborn children, whose pharynx and air-passages are filled with mucus, and in whom the removal of this mucus is one of the great desiderata.   ·

It had also frequently been recommended to lay the head over a table, to allow the head to fall back, away from the lower jaw and the structures connected with it, in order to render the mouth and upper air-passages patent.

It was decided to test this by the frozen section of a foetus in this position.

Experiment 5.—A stillborn full-time foetus which had never breathed was consequently frozen in the above position, its head being thrown as far back as it would go. It was then cut in a sagittal plane.

I have to thank Dr. Garson, of the Royal College of Surgeons, for this operation. His saw passed almost mathematically through the middle line of both the front and back of the body.

The result showed that the soft palate lay against the back of the tongue over a large extent, and did not fall away from it.

Conclusion.—Hanging the head backwards over the edge of a table does not provide for the patency of the upper air-passages.

This accords with my own clinical experience. The greatest difficulty occurs in children in the pale stage of asphyxia, and probably depends on inaction of the divari-cators of the glottis, which are flabby and useless, like all the other muscles during this stage.

This question of the patency of the upper air-passages is often the great difficulty to overcome. It frequently happens (especially in premature children, but by no means exclusively in them) that Silvester's method and its modifications entirely fail from this cause to introduce any air into the lungs. The difficulty is chiefly met with in children in the stage of pale (flabby) asphyxia, and probably for the reason mentioned above.

The method of Schultze certainly succeeds in intro-ducing air under these circumstances in many cases in which Silvester's method and its modifications have failed. The reason of this is not at present clear to me, but I have observed the fact too often to be mistaken.

I have adopted the plan (which I have not seen advocated) of introducing a catheter into the trachea and leaving it there during the respiratory manipulations, and with satisfactory results.

While in the trachea it serves for various purposes, the principal one of which is the removal of mucus, but not in the manner usually recommended. It is advantageous to secure the catheter at the proper length within the trachea, so that it may not slip too far in, or slip out, if the Schultze method is adopted.

I have measured the length along a catheter from the lips to the glottis, and find it $2\frac{1}{2}$ inches in a fœtus.

A catheter which measures $3\frac{1}{2}$ inches from the lips of the fœtus is within the glottis, but above the bifurcation of the trachea.

This then is the proper length, and it is easy to secure the catheter by a soft bandage or by a piece of elastic passed round the head, with which there is no fear of the catheter slipping out or too far in.

The removal of the mucus is best effected in the following manner, and if used should precede all attempts at respiration, in order that the mucus, meconium, and other matters found in the child's mouth may not be inhaled.

1. Lay the child on its back, with the head hanging over the edge of the table, and a little lower than the rest of its body.

2. Wipe out the mouth with a soft handkerchief.

3. Press the thorax gently with one hand, stroking the trachea upwards with the other, and retain the finger at the top of the trachea until the next manœuvre is complete. (The mucus will gravitate towards the posterior nares.)

4. Put a handkerchief over the child's mouth, blow gently, and the mucus will be blown out of the nostrils (but not into the operator's face).

5. If necessary, introduce a catheter 3½ inches from the lips and secure it.

6. Press the thorax gently with one hand (to prevent the entrance of air) and *blow* through the catheter. The opening being low down in the trachea, the air and mucus with it, being unable to pass into the lungs on account of its compression by the hand, will rush up through the glottis, and the mucus will be blown into the pharynx. This can be repeated as often as necessary, the general tendency of fluids in the air-tubes being to ascend during respiration, whether natural or artificial, towards the mouth.

This manœuvre is more efficient and far pleasanter than the suction usually recommended ; it has answered well in practice.

It must also be remembered that suction through the catheter necessitates a down-draught in the trachea, which will carry downwards any mucus, &c., within its reach.

During artificial respiration fluid will often pour from the catheter, especially if it is held over the edge of the table like a syphon.

N.B.—The plan of *holding the nostrils* during inflation of the lungs is altogether to be condemned ; it retains mucus which might be expelled, and the nostrils are a *valuable safety-valve* guarding the lungs from over-distension.

iv. I have thought it worth while to describe some of the progressive signs of returning life in a case in which I had good opportunities of observing them.

A very large male child was delivered on April 14, 1878, in a state of pale (flabby) asphyxia after turning, indicated by a hand presentation. There was exomphalos, and mouth-to-mouth inflation showed the air to pass freely into the intestines, which protruded considerably.

The child was born at 3.20 p.m., the only sign of life being a very feeble and slow beating of the heart.

Artificial respiration was begun by mouth-to-mouth inflation, and continued by the methods of Silvester and Schultze, assisted by alternate hot and cold baths. Under this treatment the heart's action became stronger and more rapid.

The first breath was drawn at 3.50 p.m., but breathing was not spontaneous till 4.30 p.m. and was then very feeble. When it was at length established, expiration was assisted by pressure on the thorax and abdomen.

The breaths were not equally drawn, but about every third breath was deeper, the intermediate breaths being light and shallow.

Breathing was fairly established at 5.15 p.m.

The first sign of muscular action was seen about 4.45 p.m., and consisted in a quivering of the tongue; then the alæ of the nose dilated, then the mouth quivered. About this time the expirations began to be tremulous as in crying. About 5 p.m. there was "sobbing," i.e. the diaphragm produced irregular inspirations without consentaneous opening of the glottis.

General Conclusions.

1. The presence of air in the abdominal viscera in no way impedes the ventilation of the lungs by artificial respiration.

2. Pressure on the cricoid cartilage, and bending the head back are powerless to prevent the passage of air into the stomach, during mouth-to-mouth inflation.

3. Hanging the head backwards over the edge of a table does not provide for the patency of the upper air-passages.

A plan of providing for the patency of the air-passages and the removal of mucus is here related in detail.

Notes of the signs of returning life in a deeply asphyxiated child are given from a careful observation.

Works quoted.

Scheel (Paul). Ueber Beschaffenheit und Nutzen des Fruchtwassers in der Luftröhre der menschlichen Früchte. Aus dem Latein, mit Anmerkungen. Erlangen, 1800, S. 53.

Herholdt (Johann Daniel). Commentation über das Leben, vorzüglich der Frucht im Menschen, und über den Tod unter der Geburt. Kopenhagen, 1803, S. 164.

Report of the Committee appointed by the Royal Medical and Chirurgical Society to investigate the subject of Suspended Animation. Med.-Chir. Trans., vol. xlv. 1862, p. 490.

Olshausen (R.). Die Behandlung scheintodter Neugeborener durch künstliche Respiration. Deutsche Klinik, 1864, Band xvi. S. 346, 357, 365.

Howard (Benjamin). Observations on the Upper Air-passages in the Anæsthetic State. Lancet, 1880, vol. i. p. 796.

Champneys (F. H.). Artificial Respiration in Stillborn Children. Med.-Chir. Trans., vol. lxiv. 1881, p. 41.

CHAPTER VI.

EXPIRATORY CERVICAL EMPHYSEMA, THAT IS, EMPHYSEMA OF
THE NECK OCCURRING DURING LABOUR AND DURING
VIOLENT EXPIRATORY EFFORTS.

(From Vol. LXVIII. of the "Medico-Chirurgical Transactions.")

IN spite of the fact that emphysema of the neck, face,
and other adjacent parts is an accident of labour which
is not extremely rare, its pathology has hitherto been the
subject of various hypotheses, amounting to little more
than guesses, and even at the present time has never,
hitherto, been the subject of accurate investigation.

The present paper is an effort to place the pathology
of this affection on a firmer basis, and the investigations
here detailed suggested themselves by way of corollary to
experiments concerning the subject of artificial respiration in newborn children and kindred subjects related in
the "Med.-Chir. Transactions," vol. lxiv. 1881, pp. 41–
101 (chapters i. and ii. of the present treatise), and vol. lxv.
1882, pp. 75–86 (chapter iii. of the present treatise). . But
the latter investigation, concerning mediastinal emphysema and pneumothorax after tracheotomy (in which the
route followed by the air was proved to be from the
tracheotomy wound, beneath the deep cervical fascia and
so into the tissue of the anterior mediastinum beneath
the pleura), suggested that emphysema of the neck

during labour might follow the same route in an opposite direction. This it was determined to test by experiment on the fœtus, for the reasons that although it is true that the affection does not concern fœtuses but adult women, yet (1) fresh fœtuses have usually fairly healthy lungs ; (2) their nearly identical age makes them a nearly uniform material ; (3) the experiments could in any case only be illustrative, and therefore could only possess value in proportion to the amount of correspondence which they showed to the known phenomena. In this way adult subjects could only serve in the same manner as fœtuses. With these considerations it was determined to choose fœtuses as the subjects of experiment. Two children only who had breathed for a very short time (Experiments 9 and 10) were also used.

Frequency of occurrence of emphysema.—Johnston and Sinclair (p. 517) had seven cases in the course of seven years (13,748 labours) in the Dublin Hospital, or less than 1 in 2,000 cases.

Ætiology.—It is generally agreed that emphysema of the neck occurring during labour is due to bearing down. This will be seen on reference to almost any case on record (see table of references). Thus in the Dublin Hospital all the patients affected were primiparæ, and in all the cases severe bearing-down efforts are noted ; it is obvious that any cause of obstruction to labour must so far favour it.

The time of its occurrence tells the same story. It never occurs before the second stage of labour, though it may not show itself till the third stage or altogether after labour (see *Dunn's* first case). In any case, however, it is no doubt produced during the second stage, even if it does not appear till later.

It need not occur, however, in connection with parturition at all, but may follow violent coughing (as in *Roché's*

case, in which there was a foreign body in the trachea),
or perhaps straining at stool (see remarks by *Dr. Otis* on
McLane's case). Indeed the bearing down in labour
has nothing special about it except that straining is maintained for a longer time than under other conditions.

Clinical course.—It will perhaps be well to quote one
or two recent cases. Those of *Dunn* and *McLane* are
selected as among the latest recorded.

Dunn records four cases of emphysema during labour;
with two of which no details are, however, given. These
are obviously cases of emphysema beginning below, and
probably in connection with rupture of the uterus. Both
ended fatally. This subject is alien to the present
inquiry. The other two cases are more to our purpose.

(1) The first of these occurred in a primipara, who had
an "ordinary labour" (whatever that may mean).
During a violent effort to expel the placenta the right
side of the face became swollen and crepitant. (N.B.—
It came out in the discussion that the swelling was first
noticed by the patient, and not till half an hour after the
expulsion of the placenta, which throws considerable
doubt on the time of the occurrence.) The swelling
seemed to begin on the lower part of the right sub-
maxillary region, and extended upwards on the nose and
cheek as far as the zygoma; it gradually extended to the
neck; no other symptoms were noticed. After some
hours, the swelling, which was puffy and moveable,
extended to the upper side of the chest and right arm.
There was no cough, pain, or dyspnoea, and no change in
the respiration. In the next two days the emphysema
extended lower on the chest. The air remained four
days and then gradually disappeared.

(2) The emphysema in the second case began (?) on
the upper part of the right side of the chest under the
clavicle, and extended upwards as far as the zygoma and

down to the middle third of the chest; also to the shoulder and upper part of the arm.

The details are so meagre that the order of the appearance of the emphysema is open to considerable doubt, but the localities affected may be accepted as correct.

McLane's case was that of a primipara, æt. 21. The first stage was very painful, the pains being long and very severe, lasting two hours, and necessitating the use of chloroform.

During the second stage there was violent straining; during one of these efforts the face became congested and purple, a swelling appeared on the right side near the trachea, and became much larger during the next three or four pains, which followed each other rapidly. The swelling extended to the right cheek. As soon as the patient recovered from the chloroform she complained of a constriction in the throat, and had some difficulty in swallowing. "Her neck, previously slender, was now thick and œdematous; her face so puffed that on the right side the eyelids were closed, as if by dropsical effusion, and the features effaced." The swelling crackled on touch. There was no emphysema below the clavicle. The whole swelling had disappeared in a week.

We learn from this and other cases that the swelling appears during the second stage of labour, generally during some violent straining effort, that it first appears in or about the suprasternal notch, extends rapidly upwards along the neck, and may reach the face; it may also extend down over the chest and down the arms. Its course reminds us of that of extravasation of urine affecting the opposite sex in a different part of the body, except that in the latter case the urine is unable to travel down the limbs. To trace the possible course of the air or the urine would be waste of time, since air easily travels all over the surface of the body in the subcu-

H

taneous cellular tissue—a fact practically useful in the skinning of animals—and urine for its part may do the same. All varieties of its extension may be found in the appended references. The other points of interest are that the air is all absorbed in a week or so, without any ill-consequences, and never ends fatally (a fact partly explaining the doubt as to its pathology).

The last, but perhaps most important, fact is that there is no disorder of respiration, and *never pneumothorax.*

Pathology.—We now come to the chief subject of our investigation.

The theories to account for the phenomena are various.

Cloquet (p. 33) gives a good account of a case which of course recovered. He says, " It would be impossible for me to define here precisely in what spot the bronchi or their branches were ruptured."

Ménière quotes Cloquet.

Blundell (p. 473) attributes it to "*rupture of the trachea and bronchi.*" He saw it twice in the same patient, a vociferous Hibernian, and in no other case.

Depaul (p. 689) gives two post-mortem accounts in which there was interlobular emphysema of the lung, but not of the neck.

Watson (vol. ii. p. 176) says, " Air passes into the mediastinum, and so into the neck."

Johnston and Sinclair do not explain it.

Roché (p. 252) in a case in which a foreign body had got into the trachea found air in the mediastinum, and double interlobular emphysema, especially on the left side.

Soyre refers to Cloquet's case, and says, " The *rupture of the trachea* was situated a little above the bifurcation of the bronchi."

Oppolzer (p. 582) says that the commonest site of interlobular emphysema is the anterior edges of the upper lobes. " Serious consequences may ensue, if the air advances towards the hilum of the lung and the mediastinum, and from thence ascends into the subcutaneous cellular tissue of the neck and face. This is, however, on the whole a rare event."

Traube (p. 89) quotes Roché.

Mackenzie (p. 205) quotes Watson.

Whitney (p. 350) relates a case, but does not speculate on the pathology.

Haultcœur (p. 420) refers to Depaul.

Schroeder (p. 455) simply says it is due to "rupture of the air-vesicles."

Prince relates a case only.

Worthington does the same.

Atthill refers to Depaul.

Alexeef refers to Depaul and Haultcœur.

Nelson relates a case.

Spiegelberg (p. 419) mentions it only as an evidence of the force of bearing down.

Dunn (p. 397) remarks that in his second case interlobular emphysema due to laceration of the air-vessels or of the bronchial tissue was possible. This rupture generally causes emphysema of the pleura, which may extend to the subcutaneous tissue of the thorax and body generally. He remarks that Jones and Sieveking say that the same condition may arise from laceration of the trachea. I have repeatedly searched for this reference but failed to find it.

McLane (p. 582) believed the lesion in his case to have been rupture of the trachea. He had seen but one other case, and in that there was rupture of the pulmonary vesicles.

We have thus a variety of theories :

(1) Rupture of the bronchi.

(2) Rupture of the trachea.

(3) Rupture of the lung.

· The *post-mortem records* bearing at all on the subject are meagre in the extreme ; they comprise two autopsies, showing interlobular emphysema of the lung, but not of the neck (Depaul), one case of interlobular emphysema, and air in the mediastinum (Roché) ; the other theories are not supported by any actual observations that I have been able to find.

It became necessary then to put these theories to the test.

Mode of experiment.—It was obviously impossible to exactly imitate bearing down, but on considering the conditions it appeared that they would be best imitated by putting pressure on the lungs filled with air and within the thorax. If any encouraging results followed it would then be advisable to study the behaviour of the lungs removed from the chest, so as to estimate the influence of the thoracic box.

The desideratum was *to produce emphysema of the neck without pneumothorax.* If pneumothorax should occur in some cases and not in others it would remain to eliminate its influence by a consideration of the cases.

Apparatus.—The apparatus consisted of a simple mercury manometer marked in millimetres, attached to a tube ending in a tracheal cannula. A T-piece answered the purpose of a mouthpiece for the operator, connected on one side with the manometer, and on the other with the trachea of the fœtus through the tracheal cannula.

Some experiments were performed with the lungs *in situ*, others with the thorax open, others with the lungs removed from the chest. In one experiment an attempt was made to trace the course of the air in former experiments by means of a coloured gelatine mass.

EXPERIMENT 1.—March 24th, 1882. Male child, born March 23rd.

Experiment March 24th, 3 p.m. (about thirty-six hours). Tracheotomy was performed, a cannula tied into the trachea and connected with a V tube filled with mercury, to the india-rubber tube of which was connected a T-piece, through which inflation was practised.

On blowing through the T-piece the chest expanded, then two small projections appeared below the third rib on each side in the nipple line; shortly afterwards the skin along the right sterno-mastoid became swollen; the swelling was tympanitic; on pressure air escaped from the tracheotomy wound.

The extreme height of the manometer = 50 mm., *i.e.* height of mercurial column = 100 mm.

Autopsy.—Thorax opened under water. Double pneumothorax. Air-bubbles beneath pulmonary pleura at the reflection of the pleura from the root of the right lung over the anterior mediastinum. Some diffuse subpleural emphysema.

On inflating the lungs again after opening the thorax, air distends the reflection of the pleura over the root of the right lung still more, and on the left side in the same manner, raising the pleura from the pericardium, and following the whole course of both phrenic nerves, dissecting up the pleura.

On plunging the fœtus again under water and blowing, air escapes from two small holes in the anterior inferior part of each lung, but in larger bubbles from the reflection of the pleura over the root of the right lung (? were the small holes in the lungs punctures or ruptures).

The region of the cervical emphysema was next dissected; the air was found along the course of the right internal jugular vein :

. ' (*a*) A bubble could be pressed into this collection from that in the anterior mediastinum.

(*b*) A bubble could be pressed the reverse way.

The air seemed to have passed from the anterior surface of the right lung—near and in front of its root—into the anterior mediastinum, behind the vena cava superior, and so along the right internal jugular vein.

Remarks.—The object of the experiment was to see in what direction air would escape from the lungs if over-blown. This succeeded, the air passing from the lung substance near the root, behind the pleura, and along the great vessels of the neck.

Pneumothorax was, however, produced on both sides.

EXPERIMENT 2.—March 31st, 1882. Fœtus born March 29th, after placenta prævia (? full time).

Experiment March 31st, 3 p.m. Tracheotomy was performed and a cannula tied into the trachea as before.

Inflation was then practised intermittently, so as to imitate, as far as possible, bearing down during labour.

The maximum height of the mercury column = 60 mm., *i.e.* whole column = 120 mm.

Autopsy.—On opening the thorax, air escaped from both pleuræ. Universal subpleural emphysema. Air occupies the anterior mediastinum and spreads in front of the root of each lung into the anterior mediastinum. No emphysema in right side of neck. Air extends along the course of the left internal jugular vein and can be pressed into the collection in front of the root of the right lung, to and fro ; it passes under the left innominate vein.

On reinflating the lungs, air is distinctly seen to pass from beneath the pleuræ near the root of the right lung, in front of the right lung into the anterior mediastinum.

Similarly on the left side, where the air chose the same path upwards along the left phrenic nerve. Air escaped from several ruptures in the pulmonary pleuræ.

Remarks.—The object and results practically the same as in Exp. 1.

EXPERIMENT 3.—April 4th, 1882. Full-time male child, born after turning, April 2nd, 2 a.m.

Experiment April 4th, 3 p.m. An incision was made from the ensiform cartilage to the pubes. Part of the right fourth rib was excised, the pleura exposed, and a cannula tied into the pleural sac, and then connected with the manometer as before.

Inflation was then practised, and the mercurial column raised to 50 mm., *i.e.* the column = 100 mm. in height; at this height air escaped through a hole made in dissection at the junction of the diaphragm and right lower ribs, wounding the diaphragm but not the pleura, but leaving the pleura unsupported, and therefore weak, at this point; from this spot the air travelled behind the peritoneum, stripping it up. The cannula was quite secure. On the left side the same condition was found.

Remarks.—The object was to compare the strength of the lung and pleura, or rather of the lung and pulmonary pleura, with that of the parietal pleura. The experiment failed, but it showed the extreme ease with which air travels beneath the pleura and peritoneum.

EXPERIMENT 4.—April 5th, 1882. Full-time male child, born April 2nd, 1882.

Experiment April 5th, at noon. The same apparatus was used as before, except that the india-rubber tube, instead of passing directly to the tracheal cannula, passed to a Wolff's bottle filled with a warm solution of coloured gelatine, which was therefore to be driven into the lungs.

On blowing at the T-piece the gelatine solution passed into the lungs.

The maximum height of the mercurial column = 30 mm., *i.e.* the height of the whole column = 60 mm.

Autopsy.—Coloured gelatine in both pleuræ. The pleuræ universally œdematous with the same fluid, but most has collected in front of the root of each lung, in the anterior mediastinum, but not along the great vessels of the neck. On blowing air through the tracheal tube, air escapes and forms emphysema on the front of the left lung in one place, and on the front of the root of the right lung, but in no other part. The most œdematous parts are perfectly air-tight.

Remarks.—The object was to produce, if possible, emphysema of the neck with some coagulable injection, to allow of a leisurely dissection. The gelatine, however, simply transuded, almost as if through a filter. I had observed this in the case of water in an experiment not here recorded. The most remarkable fact is that the lungs are simple filters to fluid when they are still absolutely air-tight. The bearing of this is wide, but need not here be enlarged upon.

EXPERIMENT 5.—April 5th, 1882. Female child, born April 13th, weight 4 lbs., length 17½ in.

N.B.—Manometer was broken at the beginning of the experiment and could not be used.

Tracheotomy, cannula tied into trachea. Lungs inflated intermittently to imitate intermittent bearing down during labour.

Autopsy.—Double pneumothorax.

General subpleural emphysema, *especially in front of both roots of the lungs;* no emphysema of the neck.

On inflating the lungs again, air slightly escaped from a few places.

Remarks.—In the absence of the manometer, an attempt was made to repeat the production of emphysema of the neck, but without success. The fœtus was probably immature.

EXPERIMENT 6.—April 17th, 1882. Male child, born April 15th, 9 a.m. (hand presented), full time. Length 20½ inches, weight 7 lbs.

Experiment April 17th, 11 a.m., *i.e.* fifty hours after death, continued for an hour and a half. Tracheotomy, cannula tied into trachea, inflation intermittent.

An escape of air from the incision was noted, ? from trachea. Cannula was tied in again.

Escape of air continued. The skin incision was prolonged to the sternum, the trachea was opened as low as possible, and the cannula tied in again.

Escape of air continued, apparently not from trachea, but from right side of it, in the region of the large vessels.

N.B.—The escape of air is only occasional, and does not prevent the mercurial column from rising to 40 mm. (*i.e.* column = 80 mm.).

Autopsy.—Double pneumothorax.

Both lungs show subpleural emphysema, the right more than the left.

A large collection of air-bubbles existed on either side in front of the root of each lung, the air having got beneath the reflection of the visceral pleura over the mediastinum nearly everywhere on the right side. Another collection of bubbles on each side where the phrenic nerves reach the diaphragm. A large collection in front of the base of the right lung. The *right* half of the anterior mediastinum is one huge bubble; there are no bubbles on the left half.

The right side of the neck was very carefully dissected.

Air is present beneath right sterno-mastoid muscle, corresponding to the place from which it escaped during the experiment. A continuous chain of air-bubbles extends from the anterior mediastinum to the collection along the vessels of the neck on the right side, following the course of the phrenic nerve. Air can be pressed from collections on the upper part of the diaphragm, along the course of each phrenic nerve to the collection in front of the root of each lung.

On the left side a few small bubbles are found in the neighbourhood of the incision, and seem to have come from the incision.

On reinflating the lungs, air can be seen to pass behind the pleural reflection into the mediastinum as above, and to escape freely from the surface of both lungs.

Remarks.—The object was still the same, and the experiment succeeded in showing plainly the route taken by the air from the lung to the neck, but pneumothorax was also produced.

EXPERIMENT 7.—June 26th, 1882. Small fœtus, obviously premature, born June 24th.

Experiment June 26th, 5 p.m. Weight of fœtus 2¾ lbs., length 15½ inches. Tracheotomy, cannula tied into trachea. Intermittent inflation. Maximum rise of mercurial column = 40 mm. (*i.e.* height of whole column = 80 mm.). Air escaped eventually from the right side of the incision.

Autopsy.—On opening the chest one or two doubtful bubbles escape from the right pleural sac, *none from the left.*

Anterior mediastinum airless in most parts. Along the great vessels on both sides (especially the right) there is a continuous chain of air-bubbles, passing behind the in-nominate vein towards a collection in front of the root of each lung.

Subpleural emphysema (slight) in one or two patches in front of each lung, but especially on their inner surface near the root.

Remarks.—The object was the same as that of the preceding experiments, and the result very instructive. On the right side the air had taken the usual route, but there was a little doubtful pneumothorax. On the left side, however, the air had passed into the neck without escaping into the pleural sac.

EXPERIMENT 8.—November 3, 1883. Male, full-time fœtus, quite fresh.

Experiment at 3 p.m. Thoracic viscera with the trachea were removed *en masse*. The lungs showed a few superficial lobules expanded, mostly at the apices.

The trachea was attached to the manometer, and a looking-glass arranged so as to show the side of the lungs away from the experimenter.

The pressure was begun at a rise of 5 mm. (*i.e.* the height of the mercury column=10 mm.), and the order of expansion of the different parts observed. It was as follows:

(1) Left apex.

(2) Right apex.

(3) Vertical strip along costal angles behind.

(4) Extension downwards over both back and front, most extensive on the right side, a few isolated patches refusing to expand.

(5) Slight bubbling below front of root of left lung.

(6) „ above „ „

(7) „ below „ right lung.

(8) „ to outer side of „ „

One patch of subpleural emphysema the size of a hemp-seed on the inner surface of the right upper lobe.

(Up to this time the rise of the mercury column did not exceed 10 mm. [*i.e.* the height of the whole column =20 mm]).

Bubbling continued all round the root of the right lung, especially below, the spot of emphysema increased, and one or two smaller ones formed near it.

Pressure was now increased to a rise of 15 mm. (*i.e.* height of whole column=30 mm.).

A large patch of emphysema formed at once on the inner side of each upper lobe, as large as half a marble.

(N.B.—These patches spread when the pressure only causes a rise of 5 mm., *i.e.*, under a column of 10 mm.)

One or two very small patches of emphysema were seen at the sides of the lungs, but none in front.

A large patch formed on the under surface of the right upper lobe, also in the fissures of the right lung.

The left lower lobe was then treated separately, the cannula was thrust into the lung substance and secured, the lung was then inflated. Immediately, a large bubble rose on the surface, disappearing as soon as pressure was relaxed, and reappearing instantly when it was resumed. It was found impossible to burst this even when 100 mm. pressure (*i.e.* a rise of 50 mm.) was steadily and repeatedly produced.

The bronchi being tied, it was found impossible to burst them under the greatest expiratory force available (=a rise of 50 mm., or a column of a height=100 mm.).

Remarks.—The object was to see the behaviour of the lungs themselves uninfluenced by the thorax, (1) with regard to the order of expansion of different parts, (2) with regard to the comparative tenacity of different parts of the lungs, (3) of the pulmonary pleura, (4) of the bronchi.

(1) The order of expansion is given above. It is curious that the vertical strip along the costal angles, which is one of the last places to be expanded when the lungs are within the thorax, should have been one of the first to expand here.

(2) The weakest part of the lung was seen to be about its root, which gave way (subpleural emphysema) under the pressure of a column=20 mm.

(3) The strength of the pulmonary pleura was seen to vary immensely; here it could not be ruptured, in former experiments the pleura was ruptured by far less pressure.

(4) The bronchi could not be ruptured by the strongest expiratory effort.

Emphysema, once formed, was seen to spread at a comparatively small pressure (column=10 mm.).

EXPERIMENT 9.—November 8th, 1882. Fœtus female, born November 6–7 (midnight) and lived eight hours. Artificial respiration had been performed.

Experiment November 8th, 10 p.m. Thorax opened, but lungs not removed. No pneumothorax; a few small subpleural ecchymoses, lungs partly expanded.

Tracheotomy, cannula tied in trachea. Inflation (to a rise=10 mm., or a column=20 mm.) caused no emphysema.

The pressure was increased to a rise of 15 mm. (or a column = 30 mm.). Slight scattered interlobular subpleural emphysema followed, most marked in the left lung.

On repeating inflation under the same pressure, emphysema spreads, especially in the fissures.

The pressure was increased to a rise of 20 mm. (or a column=40 mm.), and a bubble of subpleural emphysema on the inner side of the right middle lobe gave way.

After several repetitions, a large bubble was seen to

occupy the posterior mediastinum, extending below and in front of the root of the left lung, and gradually extending .upwards thence into the anterior mediastinum between the thymus and pericardium.

Although pressure was repeatedly increased to a rise of 25 mm. (or a column=50 mm.) and although the posterior mediastinum was full of air as far as the diaphragm, no air rose into the neck.

Remarks.—The object was to observe the behaviour of the lungs *in sitû*, but with the thorax opened.

(1) No emphysema took place till a rise of 15 mm. (or a column=30 mm.), *i.e.* the lung substance gave way at this pressure.

(2) The first emphysema was between the lobules at the surface of the lung.

(3) Emphysema appeared very early between the lobes.

(4) The pleura gave way at a rise of 20 mm. (or a column = 40 mm.).

(5) The air did not rise into the neck.

EXPERIMENT 10.—November 9th, 1882. Male, seven months child, born midnight November 6th–7th, died 12.30 p.m. November 8th (thirty-six hours).

Thorax opened. Tracheotomy, cannula in trachea.

On inflating the lungs, a leak is seen in the inner edge of the left upper lobe (the lung had probably been pricked in opening the chest). The leak cannot be stopped.

Experiment failed.

EXPERIMENT 11.—December 29th, 1882. Full-time stillborn male child, of somewhat doubtful freshness (date of birth unknown).

Thorax opened, cannula tied into trachea.

Lungs entirely airless, very watery.

Slight expiration inflates patches of the lungs, especially on the posterior surface.

Additional inflation produces increased expansion of the same parts, least behind.

Eventually the lungs became almost entirely expanded. The pressure was increased to a rise of 10 mm. (or a column = 20 mm.). No emphysema.

The pressure was increased to a rise of 15 mm. (or a column = 30 mm.). Considerable emphysema of the front of the root of the right lung occurred, extending into the fissure in front between the right upper and middle lobes, and also (but less) emphysema of the anterior and inner aspects of the upper and middle lobes. It began most markedly between the lobules. A bubble burst on the inner and anterior surface of the right upper and middle lobes.

Numerous leaks having occurred through the pleura the root of the right lung was tied.

Pressure producing a rise of 10 to 15 mm. (or a column = 20 to 30 mm.) produced well-marked emphysema of the front of the root of the left lung, extending in the anterior mediastinum and ⅛ inch along the course of the left phrenic nerve.

On repeating this, the inner surface of the upper lobe became emphysematous, and a bubble in that situation burst. Emphysema extended into the fissure between the upper and middle lobes.

The pressure was increased to a rise of 20 mm. (or a column = 40 mm.), and produced much escape of air from subpleural bubbles and extension of the emphysema from the root of the lung downwards into the posterior mediastinum.

The *bronchi were then tied*, and the greatest possible expiratory force exerted, causing a rise of 75 mm. (or a column = 150 mm.).

No escape óf air took place.

Remarks.—(1) Emphysema (= rupture of the lung) tóok place at a pressure = a column 30 mm. high.

(2) The pleura eventually gave way at this pressure.

(3) The front of the root of each lung was the first place to give way.

(4) The spaces between the lobules and lobes (*i.e.* the interlobular spaces and fissures) were weak places.

(5) The bronchi and trachea could not be ruptured by the greatest expiratory effort.

EXPERIMENT 12.—December 30th, 1882. Large full-time female child, stillborn (placenta prævia), about twenty-four hours.

Thorax opened; lungs unexpanded and very sodden.

Tracheotomy, cannula tied into trachea. Inflation was begun at a rise of 10 mm. (or a column = 20 mm.), and produced scattered inflated patches on the anterior and internal edges of all the lobes of both lungs, most marked inferiorly.

Second inflation at the same pressure produced an increase of the same, plus slight subpleural emphysema of the left base posteriorly.

The third inflation at the same pressure fully inflated the lungs and slightly increased the emphysema.

The pressure was increased to a rise of 15 mm. (or a column = 30 mm.), and produced a leakage of air from the left base.

The left base, seeming to be unusually frail, was tied off, and the experiment resumed.

On repeating the inflation at the same pressure a large patch of emphysema occupied the tip of the lingula. A large bubble formed in front of the root of the left lung, and smaller ones on the inner surface of the left upper lobe, mostly interlobular. The emphysema, still mostly interlobular, extended over the left lung.

The left lung was next tied at the root, and the experiment continued with the right lung.

Inflation was continued with a rise of 15 mm. (or a column = 30 mm.), which produced interlobular emphysema of the right base. Large bubbles occupied the fissures between the middle and lower lobes.

The pressure was increased to a rise of 20 mm. (or a column = 40 mm.); the emphysema extended over the front of the middle lobe, and a bubble burst, probably in one of the fissures.

The root of the right lung was tied and the strongest expiratory force exerted. No leak was produced at a rise of 80 mm. (or a column = 160 mm.).

Remarks.—The order of inflation was unlike the former experiments, perhaps in consequence of the sodden state of the lungs.

(1) The left lung gave way (emphysema) under a column 20 mm. high, the right under a column 30 mm. high.

(2) The left pleura gave way under a column 30 mm. high, the right under a column 40 mm. high.

(3) The spaces between the lobules and lobes (interlobular spaces and fissures) were weak places.

(4) The bronchi and trachea could not be burst by the greatest expiratory effort.

EXPERIMENT 13.—January 1st, 1883. Full-time stillborn child (craniotomy), born twenty-four hours previously.

Thorax not opened. Tracheotomy, cannula tied into trachea and attached to mercury manometer as before.

The lungs were inflated at a rise not exceeding 10 mm. (or a column 20 mm. high), then the thorax was compressed (the escape of air being prevented), causing an increased rise of 30 mm. (or a column 60 mm. high).

I

This was several times repeated, to imitate as far as possible the effect of bearing down.

A projection is seen above each clavicle, apparently from the apices of the lungs.

After several repetitions the left side of the neck was seen to be full, and on pouring water on the tracheotomy incision bubbles were seen to escape from the left side of the wound.

The initial pressure was now increased, so as to produce a rise of 15 mm. (or a column = 30 mm.), and then a rise of 20 mm. (or a column = 40 mm.).

Autopsy.—Considerable emphysema was found round the root of left lung, extending into the fissures and posterior mediastinum. Interlobular emphysema of whole of left lung, air escaping from several places. Slight diffused subpleural and interlobular emphysema of right lung, especially in front of its root.

No emphysema of neck.

The bronchi being tied, the greatest possible expiratory efforts, producing a rise of 80 mm. (or a column = 160 mm.), produced no escape of air.

Remarks.—(1) The dissection threw doubt on the escape of air from the neck, at least no air was found along the great vessels on dissection.

(2) The emphysema was seen to occupy the interlobular spaces, fissures, and the front of the root of the right lung.

(3) The bronchi and trachea could not be burst by the greatest expiratory effort.

Experiment 14.—February 22nd, 1883. Full-time stillborn male child (face presentation), born February 19th.

Experiment February 22nd, noon. Thorax not opened. Tracheotomy, cannula tied into trachea.

The lungs were then inflated by intermittent inflations,

the pressure causing a rise not exceeding 15 mm. (or a column = 30 mm. high), and these inflations were continued half an hour.

Autopsy.—Lungs completely inflated, no emphysema. Inflation was then continued at same pressure; no further result.

Pressure was then increased so as to produce a rise of 20 mm. (or a column = 40 mm.), producing a small spot of *subpleural emphysema* on the anterior surface of the internal inferior angle of the right middle lobe, and in several *interlobular* spaces on the inner surface of the right upper and middle lobes.

On repeating the inflation, the emphysema spread and extended in front of the root of the right lung, from which air then escaped.

The emphysema extended to the left lung and in front of its root, markedly between the lobules and in the fissures.

The emphysema then extended slightly into the right side of the *anterior mediastinum*, above and in front (from the front of the root of the right lung), and into the left side of the posterior mediastinum behind and below.

The emphysema then occupied all the fissures of both lungs.

The right side of the posterior mediastinum then became emphysematous.

Both lungs eventually showed great emphysema behind all pleural reflections.

Remarks.—The lungs were unusually air-tight; it was not till a rise of 20 mm. (or a column of 40 mm.) was produced that any emphysema appeared.

(1) The lung tissue gave way at a rise of 20 mm. (or a column of 40 mm.).

(2) The pleura gave way at the same pressure when prolonged.

(3) The pleural reflections and the interlobular spaces were the weakest. (These include the fissures and the reflection of the pleura over the root of the lung to the anterior mediastinum.)

EXPERIMENT 15.—February 27th, 1883. Male stillborn child (premature labour at end of eighth month), born February 26th, 2 p.m.

Experiment February 27th, noon.

Tracheotomy, cannula tied into trachea. Lungs were inflated at a pressure causing a rise not exceeding 15 mm. (or a column = 30 mm.). This was repeated intermittently for fifteen minutes, when the pressure was increased to a rise of 20 mm. (or a column = 40 mm.).

No signs of emphysema of the neck.

Autopsy.—No emphysema except a small spot on the inferior edge of the front of the right middle lobe (? produced in opening thorax), from which air escapes. On holding this, the emphysema spreads widely over all surfaces of the right lung. The right lung was then tied at its root. The left lung sustained a pressure causing a rise of 20 mm. (or a column = 40 mm.). On increasing the pressure to a rise of 40 mm. (or a column = 80 mm.), diffuse subpleural emphysema formed over posterior and inferior surface of back of left base.

Remarks.—The lung tissue only yielded to a pressure equal to a column of mercury 80 mm. high.

EXPERIMENT 16.—March 17th, 1883. Stillborn male child, born evening of March 15th.

Experiment at 11 a.m., March 17th. Thorax not opened. Tracheotomy, cannula tied into trachea.

The lungs were inflated intermittently, the pressure being gradually increased from a rise of 15 mm. (or a column = 30 mm.) to a rise of 40 mm. (or a column = 80 mm.), at which pressure the column sank as if from a leak.

Autopsy.—Air in both pleuræ, a good deal of fluid; lungs very œdematous. One or two very small patches of subpleural emphysema on both lungs. In front of the root of the right lung is a large patch extending into the anterior mediastinum in front, into the posterior mediastinum (which is greatly distended) behind, and into the fissure between the right middle and lower lobes.

On the left side the root of the lung is simply surrounded with emphysema, raising the reflection of the visceral and parietal pleura along the inner edge of the lung in its whole length, thus filling the posterior and left half of the anterior mediastinum. From the front of the root of the right lung bubbles extend to the phrenic nerve and along its course to the diaphragm, where there is a large collection of bubbles. No collection of bubbles is seen on dissection on either side of the neck, but they can be pressed from the anterior mediastinum upwards and escape alongside of the great vessels on the left side, passing behind the left innominate vein.

Summary.—Rupture of both lungs about their root at a pressure of 80 mm. of mercury; escape of air into both mediastina, on the left side following the phrenic nerve to the diaphragm. Pneumothorax (double), probably from escape of air from a rupture of the mediastinum.

Remarks.—The weak parts were the reflections of the pleura as before, especially the large ones in front of the root of the lungs and in the fissures.

Although air was not actually found in the neck, a free communication between the neck and the anterior mediastinum was demonstrated. The passage of air along the phrenic nerve to the diaphragm has already been noted in other cases and also in the experiments on

artificial respiration (" Med.-Chir. Trans.," vol. lxv. 1882,
pp. 77 and 80; pp. 70 and 73 of the present treatise).

EXPERIMENT 17.—March 20th, 1883. Full-time stillborn male child (second of twins) born March 19th, 3 a.m.

Experiment March 20th, noon. Thorax not opened.
Tracheotomy, cannula tied into trachea.

Lungs were inflated at a pressure producing a rise not
exceeding 15 mm. (or column = 30 mm.).

Pressure was gradually increased to a rise not exceeding
20 mm. (or column = 40 mm.), giving twenty inflations
at each rise of 5 mm. (or increase of column = 10 mm.).

On reaching this pressure, the *left side of the neck
became distended, and air escaped from the left side of the
incision.*

Autopsy.—No pneumothorax. The lungs on being
inflated after opening the thorax are found perfectly air-
tight.

Left side.—Air extends along the large vessels of the
neck on the left side, and can be traced behind the left
innominate vein to the anterior mediastinum, where a large
collection is continuous with one in front of the root of the
left lung and extending into the fissures of the lung. Air
distends the posterior mediastinum and runs beneath the
whole reflection of the visceral into the parietal pleura,
along the inner edge of the left lung, and from thence
along the diaphragm to the termination of the phrenic
nerve; a few scattered spots of subpleural emphysema,
especially on the " lingula."

Right side.—No emphysema of neck. Emphysema in
front of root of lung and upwards a little way along
phrenic nerve towards neck (N.B.—The middle lobe was
not separated from the upper) and along the inner side of
the middle lobe following an intralobar fissure; also (as
on left side) in posterior mediastinum, extending forwards

to the ending of the phrenic nerve in the diaphragm, along the reflection of the pleura over the adjacent sides of the anterior mediastinum, diaphragm, and pericardium; also upwards along the phrenic nerves.

No other emphysema.

Remarks.—Emphysema of the left side of the neck was produced without pneumothorax. The air as usual had escaped from the front of the left lung near its root, into the mediastinum and along the great vessels to the neck. The fissures were also occupied by emphysema.

The pleura (though not the lung) withstood a pressure producing a rise = 20 mm. (or a column = 40 mm.).

Consideration of the experiments.—The production of emphysema of the neck was effected in the very first experiment, but it was not till Exp. 7 that it was produced without pneumothorax (left side). This result was also achieved in the last experiment, No. 17 (left side).

The difficulty was to find the necessary pressure.

The question of the significance of the cases in which pneumothorax was produced will be discussed hereafter.

The *route selected by the air* was exactly the same as in emphysema of the mediastinum after tracheotomy (after entering the mediastinum), but in a contrary direction.

The *source of the air* was rupture of the lung tissue producing interlobular emphysema near the anterior aspect of the root of the lung.

The question *why this spot is especially prone to rupture* was capable of two answers: (1) it might be due to inherent weakness in this part of the lung; (2) it might be due to the relation between the lung and the thoracic box.

This question was investigated by experiments on the lungs with the chest open, or after their removal from the chest. The results obtained showed that in lungs removed from the pressure of the chest wall, the weakest parts are the spaces between the lobules and lobes.

- This is easily intelligible when it is remembered that the pleura invests the lung with hardly any adhesion to its surface, and that it can be stripped up with almost inappreciable force by air beneath it. The strength of the pleura is simply that of an independent investing elastic bag. If the air escapes from a lobule it finds no resistance from the pleura except on the surface of the lung, and may easily lie between the lobules without any pressure from the pleura.

With regard to the fissures of the lung, the same is true; the pleura is here, so to speak, slack and offers little or no resistance to air once escaped from the air-cells.

But experiments showed that the root of the lung was a specially weak spot with the thorax closed, though not with the thorax open. In other words, when the thorax was open it merely shared the weakness common to all the pleural reflections.

This point requires some discussion. What is the physical condition of the lung during inflation within the closed thoracic walls? The thorax is distended, the diaphragm depressed, the sternum, clavicles, and ribs elevated.

Is the thorax or the lung the more distensible? To this it must be answered that the lungs can easily be ruptured within the thorax.

But it nevertheless cannot be doubted that, up to this point, the thorax supports its contents.

Does it do so equally in all directions? Below, we have the unbroken plane of the diaphragm; behind, in front, and at the outer side we have the ribs and muscles. But at the inner side we have the compressible mediastinum, whose easy penetrability was proved in the experiments already alluded to dealing with the subject of mediastinal emphysema after tracheotomy. This side then seems to be the direction of least resistance.

When once in the mediastinum, the air is already within the track of easy penetrability or of slight resistance, leading into the neck, and which probably owes this quality partly to the fact that the upper aperture of the thorax is the weakest spot in the thoracic box, not being directly defended by muscles, which pass obliquely from the neck to the upper ribs, clavicles and sternum. It must also be remembered that this upper aperture of the thorax is enlarged on inspiration.

The above reasons show that the mediastinum may be considered to be within the area of diminished resistance.

What happens in a bearing-down effort? First a deep inspiration, which (among other things) raises the upper ribs and clavicles, and increases the size of the upper aperture of the thorax. Next the glottis is closed. Lastly the whole of the expiratory muscles, essential and accessory, put forth their strength.

The least resistance is offered to pressure at the upper aperture of the thorax.

Thus then, on the *anterior surface of the root of the lung is the pleural reflection least supported externally.*

We have now to consider the question of the significance of the occurrence of pneumothorax in some of the experiments.

Is the pneumothorax in these cases a link in the chain ending in emphysema of the neck?

A very little consideration will suffice to put such an idea aside.

Let us consider the course of the air on this hypothesis. It escapes from the air-vesicles beneath the pleura; bursts through the pulmonary pleura; must then distend the pleural cavity; then bursts through the parietal pleura, and so gets beneath the deep cervical fascia.

First of all, the autopsies entirely contradict such an assertion; they show the course of the air to be different. Secondly, the air beneath the pulmonary pleura finds practically no resistance in travelling beneath the pleura, but the pulmonary pleura is rather tougher than the lung. The experiments show the tenacity of both lung and pleura to vary very greatly in the fœtus (whatever they do in the adult), as the following table shows :—

Lowest force required to burst the lung :—

Experiment No. 8   .   .   20 mm.
        „   „ 12   .   .   20   „

Lowest force required to burst the pleura :—

Experiment No. 11   .   .   30 mm.
        „   „ 12   .   .   30   „

Highest force required to burst the lung :—

Experiment No. 15   .   .   80 mm.

Highest force required to burst the pleura :—

Experiment No. 8 (N.B., locally) 100 mm.
        „   „ 9   .   .   40   „
        „   „ 14   .   .   40   „

An attempt was made (Exp. 3) to test the tenacity of the parietal pleura, but it was unsuccessful. It must, however, be remembered that the tenacity of a membrane like the pleura must be very different when raised from the subjacent structures (pulmonary pleura), and when subjected to force, which only presses it the more firmly on its supports. It can easily be imagined that a film of collodion might add great strength to a membrane if force was applied in such a direction as to press the film against the membrane, while its strength would probably be very small if it had to resist a force from below, that is, raising it from its supporting membrane.

We therefore conclude that pneumothorax, when it occurred, had nothing to do with the production of emphysema of the neck.

We have avoided speaking of the clinical phenomena which form the actual subject of our consideration, but it must not be forgotten that pneumothorax never occurs either with emphysema of the neck during labour, or as the result of expiratory efforts.

We have not hitherto spoken of the theories that regard emphysema of the neck as due to rupture of the trachea or of the bronchi.

This may be dismissed in a word : *The strongest expiratory effort failed to burst the trachea or bronchi of a fœtus.*

The experiment was repeatedly tried, and amounts to a " reductio ad absurdum " of the hypothesis.

It cannot be alleged that in the cases in which this accident occurs during labour the patients have had disease or fistula of their bronchi or trachea, for the patients may in all cases be, and are nearly always expressly described as, perfectly healthy, and moreover they suffer no ill-effects.

To the question whether a lung can be ruptured with impunity we must answer in the affirmative; if the air is all absorbed within a week from the subcutaneous cellular tissue, why not from beneath the pleura ?

Again, the collection beneath the pleura must be small or it would give physical signs, or at least marked symptoms, which, however, are always absent.

What actually happens is probably as follows : During a violent expiratory effort an air-cell near the front of the root of the lung gives way, and the air lies beneath the pulmonary pleura. With the next effort this becomes larger, and part of it moves in the direction of least resistance, namely, towards the mediastinum, next time towards the neck, and so on, until a bubble emerges beneath the deep cervical fascia. A channel will thus be formed along which bubbles will pass as quickly almost as they escape from the lung. They may form a large collection in the neck, beneath the deep cervical fascia where the pressure is small; eventually they may find their way into the superficial fascia, and so all over the body.

The gelatine injection (Exp. No. 4) gave us the interesting fact that a lung which is quite air-tight is quite permeable for fluids. I had previously proved this as regards water, and it is seen to be true also of a colloid mass. This fact has various and important bearings, which this is not the place to enlarge upon. It may, however, be remarked that the air-cells are lined with an epithelium not much removed from an endothelium, such as that which lines serous cavities which are lymph-sacs and highly permeable to fluids.

The experiments are thus seen to illustrate the clinical facts of emphysema of the neck during labour, and during violent expiratory efforts.

The emphysema is essentially *expiratory* in its nature, and due to a cause entirely opposite to that which is answerable for emphysema of the mediastinum after tracheotomy. In saying this it is conceivably possible that expiratory emphysema might occur after tracheotomy, though the conditions under which the operation is performed are well known to obstruct inspiration rather than expiration. In this case the operation would have nothing to do with the emphysema. Emphysema of the lung may occur from over-distension of a part due to obstruction in other parts. This is a well-known clinical fact, and was also illustrated in the experiments quoted above ("Med.-Chir. Trans.," vol. lxv. 1882, p. 78, Exp. 19 (L); and p. 71 of the present treatise). In such a case the air which had escaped beneath the pulmonary pleura might conceivably be forced into the mediastinum. But the burden of proof in any case of tracheotomy rests with the observer who asserts that the emphysema which he finds, is due to expiratory rather than inspiratory causes. The mediastinal emphysema of tracheotomy is *inspiratory;* the mediastinal emphysema of violent expiratory efforts is *expiratory.* Mediastinal emphysema without subpleural emphysema cannot be expiratory. Subpleural emphysema without mediastinal emphysema may be inspiratory as well as expiratory. Subpleural, together with mediastinal, emphysema is probably altogether expiratory, but the former may be due to inspiratory over-distension of part of the lung with obstruction elsewhere; the latter may conceivably be due to the expiratory forcing of the air thus escaped into the mediastinum. We are now only speaking of cases of tracheotomy in which the derivation of the air from a cervical wound is a possibility, and the source of the air is therefore so far debateable.

But while allowing the above possibilities, we wish to repeat that the emphysema of tracheotomy occurs under conditions which obstruct inspiration, and is therefore essentially *inspiratory,* the air being derived from the cervical wound.

The following conclusions are offered :

1. The cause of emphysema of the neck during labour is rupture of the lung tissue, the air escaping near the root of the lung, passing beneath the pulmonary pleura into the anterior mediastinum, and so beneath the deep cervical fascia into the neck. The route thus marked is the same by which air sometimes passes into the anterior mediastinum after tracheotomy (see " Med.-Chir. Trans.," vol. lxv. 1882, p. 75 *et seq.,* and p. 85 ; and chapter iii. of the present treatise, p. 79).

2. The weakest parts of the lung are opposite the

pleural reflections (that is the fissures) and the interlobular spaces. The anterior surface of the root of the lung is the weakest spot while the lungs are within the thorax, being that pleural reflection lying within the comparatively unsupported area near the upper aperture of the thorax.

3. Pneumothorax, when it occurred during experiment, had nothing to do with the production of emphysema of the neck, and in two experiments was not associated with this emphysema, which thus exactly imitated that occurring during labour.

4. The healthy bronchi and trachea are able to resist the greatest possible expiratory efforts.·

5. The lungs and pleuræ when quite air-tight are freely permeable to fluids.

6. The usual rules of practice to restrain bearing down and accelerate labour after the production of emphysema of the neck are sound.

7. The accident would seem to be noted in about 1 case in 2,000, but it is not improbable that slight cases are overlooked.

8. The air emerges from the thorax along the great vessels, but may not become superficial till it has travelled higher up.

9. The emphysema of the lower part of the trunk, usually connected with rupture of the uterus, belongs to quite a different category and is generally associated with a fatal result.

List of works quoted.

Cloquet (Jules).   De l'influence des Efforts sur les organes dans la cavité thoracique.   Paris, 1820.

Ménière (P.).   Arch. gén. de Méd., xix., 1829, p. 341.

Blundell (James).   Principles and Practice of Obstetricy.   London, 1834.

Depaul. Gaz. méd. de Paris, Oct. 29, 1842, p. 689.

Watson (Sir Thomas). Principles and Practice of Physic. London, 1857.

Sinclair (Edward B.), and *Johnston (George).* Practical Midwifery. London, 1858.

Roché (L.). Bull. de la Soc. Anat. de Paris, 2e série, iv., 1859, p. 252.

de Soyre (Jules). Gaz. des Hôp., 1864, No. 92, p. 366, and No. 100, p. 398.

Oppolzer. Vorlesungen über specielle Pathologie und Therapie. Erlangen, 1866–72.

Traube (Ludwig). Die Symptome der Krankheiten des Respirations- und Circulations-Apparats. Berlin, 1867.

Mackenzie (Colin). Amer. Jour. of Obst., vol. iv., 1871, p. 203.

Whitney (James O.). Bost. Med. Surg. Jour., Nov. 30th, 1871, p. 350.

Haultcœur. Gaz. Obst., 1874, p. 420.

Schroeder (Karl). Lehrb. der Geburtshülfe, 4te Aufl. Bonn, 1874.

Prince (A.). Lancet, Jan. 15th, 1876, p. 117.

Worthington (Francis). Brit. Med. Jour., Jan. 29th, 1876, p. 124.

Atthill (Blennerhassett). Obst. Jour., vol. iv., 1876, p. 18.

Alexeef. Arch. f. Gyn. Band ix., 1876, S. 437.

Nelson (H. S.). Edin. Med. Journ., July, 1877, p. 43.

Spiegelberg (Otto). Lehrb. der Geburtshülfe, 1878.

Champneys (F. H.). Med.-Chir. Trans., vol. lxiv., 1881, pp. 41—101.

Champneys (F. H.). Med.-Chir. Trans., vol. lxv., 1882, pp. 75—86.

Dunn (W. A.). Bost. Med. Surg. Jour., April 26th, 1883, p. 397.

McLane (J. W.). New York Med. Jour., May 2Cth, 1883, p. 582.

- [See also *Chahbazian.* Arch. de Tocologie, Juillet, 1883.]

[For discussion on this paper see 'Proceedings of the Royal Medical and Chirurgical Society,' New Series, vol. i., p. 285.]

ADDENDUM.—Since the above was written, it has been confirmed by a valuable observation of Dr. Robert Boxall, now Physician to the General Lying-in Hospital, recorded in the *Lancet*, Jan. 15, 1887, p. 121, which is as follows :—

Mrs. S——, aged nineteen, a well-developed but somewhat delicate-looking primipara, was admitted towards the end of the first stage of labour on August 26th, 1886. Full dilatation of the cervix was reached at 10.20 a.m., and the head passed slowly through the pelvis in the first cranial position. It remained two hours on the perineum, and was eventually expelled without artificial aid at 1.50 p.m. The labour pains were very strong, and towards the end became almost continuous. The patient meanwhile screamed violently. The child, when born, weighed 6¼ lbs. The placenta was expressed a quarter of an hour after the birth of the child. The uterus subsequently contracted well.

About three hours after delivery the patient herself was aware of a soreness in the neck and upper part of the chest. It began about the left sterno-clavicular articulation, and thence spread upwards and to the opposite side of the neck. Next morning (27th) a puffy swelling was observed at the root of the neck over the manubrium sterni; it was a little tender on pressure. The skin retained its natural colour. At the apex of both lungs, and especially the left, the percussion note was hyper-resonant, and the breath sounds feeble. The patient had a slight

cough, but no disturbance of breathing. The heart sounds
were normal, but at the base were marked by crackling.

On August 29th, during the physician's visit, the following notes were made :—Slight fulness is apparent
about the lower part of the neck on either side, but
especially the left. The respiratory movements are rather
less marked at the left than at the right apex. Over the
puffy area emphysematous crackling is easily distinguished
by the finger. It is most marked over the manubrium
sterni, but does not extend below its junction with the
body of the sternum. It can be felt on either side immediately below the clavicle, extending as low as the upper
border of the second rib and outwards as far as the coracoid
process. It is most prominent on the left side, extending
backwards as far as the edge of the trapezius (where it
appears to end abruptly) and upwards to within an inch
of the mastoid process, but is less distinct, and only
brought out by deep palpation. In the anterior triangle
it again becomes distinct, extending upwards quite as far
as the ramus of the jaw and across the middle line to the
opposite side. On the right side of the neck it is as
distinctly felt in the anterior triangle as on the left side,
but cannot be felt in the posterior triangle. No crepitation can be felt under the trapezius, over the mamma, or
down the part of the sternum below the junction of the
manubrium and body. Over both anterior and posterior
triangles on the left side, and over the anterior on the
right, a high-pitched hyper-resonant note is easily produced. The percussion note below the clavicle on the
left is slightly higher pitched than on the opposite side.
Over the whole area above described superficial crepitation sounds are audible. These are especially numerous
on first applying the stethoscope to the surface, and are
readily brought out by shifting its position, and are produced, though to a less degree, by the inspiratory move-

K

ments. In addition, over the pulmonary cartilage they accompany each systole of the heart. No corresponding sound is heard over the aortic cartilage, though no emphysematous crackling can be felt; nor can any sounds be produced by pressure of the stethoscope below the junction of the manubrium with the body of the sternum; crepitation sounds are audible, synchronous both with each inspiration and with each systole of the heart, and extending as far down as the xiphi-sternal articulation, on the right side to the border of the sternum, and on the left a finger's breadth beyond it. In the neck, over the whole area affected, slight pectoriloquy is audible, and the voice assumes somewhat an ægophonic character. The skin still preserves its natural colour. The soreness is diminishing.

The crackling sounds accompanying the systole of the heart had disappeared by September 2nd (7th day), and all signs of emphysema by September 3rd (8th day).

No special treatment was adopted. The temperature remained normal throughout; the pulse varied from 72 to 80, and the respirations from 20 to 24 per minute.

This observation I believe to be the first recorded; others will doubtless follow.

Dr. Boxall remarks : " It seems highly probable that the air escaping at the root of the lung found its way into the anterior mediastinum; hence the crepitation sounds, synchronous with both inspiration and with the heart's systole, audible over the front of the sternum, where no air had escaped into the tissues superficial to the bone.

The air travelling upwards along the pulmonary artery would give rise to similar sounds, heard over the pulmonary cartilage with each systole of the heart; thence travelling onwards and directed upwards by the deep cervical fascia, the air entered both the anterior triangles of the neck, and on the left side (where it commenced, and which throughout was more affected than the right), passing beneath the sterno-mastoid, inflated the posterior triangle, and on either side passed over the clavicle as low as the upper border of the second rib and corresponding part of the sternum. Subsequently a little air appears to have found its way also beneath the right sterno-mastoid into the posterior triangle of the same side.

CHAPTER VII.

SOME POINTS IN THE PRACTICE OF ARTIFICIAL RESPIRATION
IN CASES OF STILLBIRTH AND OF APPARENT DEATH AFTER
TRACHEOTOMY.

(From the "International Journal of Medical Sciences," April, 1886.)

THE subject of artificial respiration is naturally preceded by a few words on stillbirth. That the popular use of "stillbirth" as synonymous with "deadbirth" is incorrect, is proved by the generally used and quite correct phrase "resuscitation of stillborn children." If the children are dead, they obviously cannot be resuscitated. Again, the phrase cannot mean birth without any of the signs of life, for, speaking generally, the heart is beating. The expression, then, practically comes to mean birth without obvious movements, whether of respiration or of other kinds. This, after all, is the plain meaning of the word "still." All deadborn children are therefore stillborn, but all stillborn children are not deadborn.

A child born alive but "still"—that is, generally, but not necessarily, with its heart beating, but without movement—may be in one of two stages or states, for the description of which we are indebted to Cazeaux (loc. cit.). In the first, which he calls the "apoplectic state," the surface is livid, but the muscular tone is not lost, and there is response to reflex irritation. In the second, "syncope," the surface is pale, the muscular tone is lost, and there is no response to reflex irritation. These two states are also known as the livid stage and the pale or flabby stage of asphyxia.

The relation of these two stages to each other is not finally determined. It is known that the head of the foetus is subjected to great pressure in its passage through the genital canal, and that hæmorrhages into the brain and spinal cord are so commonly (Litzmann, *loc. cit.*) found in the bodies of children born dead or dying soon after birth, that it is probable that such hæmorrhages are by no means invariably fatal. No necessary connection apparently exists between serious lesions of this sort and stillbirth, in the sense that in cases of dangerous stillbirth such lesions are necessarily present. Moreover, the worst cases of both varieties occur where the head is born last, and has, therefore, suffered less pressure than usual.

The state of asphyxia at birth is an exaggeration of the state of strong "necessity of breathing," which is normally produced during labour. As the intrauterine pressure rises, and the placenta is pressed upon with increasing force, and as the placental site diminishes with uterine retraction, the foetal circulation becomes progressively embarrassed, the external sign of which is retardation of the foetal heart. The complete or virtual abolition of the placental circulation which generally ensues on the completion of the birth of the child, and the shock of the relatively cold air, may either of them be competent to excite the first inspiration. Sometimes, however, matters have gone further, and the child is too asphyxiated for the moment to be roused. On the other hand, pressure on the cord, or separation or great squeezing of the placenta (such as occurs in head-last cases, when the head is in the vagina, and the uterus is already in the third stage of labour before the child is born) may excite premature inspiratory efforts, convulsive in their nature, which may draw any matters in the genital passages, such as liquor amnii, vernix caseosa, and meconium, not only into the trachea, bronchi, and air cells, but even beneath the pleura. In such a case the child, resuscitated it may be, dies of a low form of lobular pneumonia.

The length of survival of apparently dead children is sometimes remarkable, as the following instances, in which children have been buried apparently dead, and afterwards dug up and restored to life, show. At a discussion on one of the author's papers, Dr. Roper related that in the practice of the late Mr. Brown, of St. Mary Axe, a child was dissected the day after stillbirth by the late Mr. Solly, and the heart found beating. Also that a child born at five and a half months and set aside for dead, having lain on the floor eleven hours through a cold night, was found breathing and its heart beating (*Lancet*, Nov. 27th, 1880, p. 852). Two illegitimate children of separate mothers were buried, and restored to life after several hours' burial. A child was buried five hours, restored to life, and lived three days (Bohn, *loc. cit.*). A child was buried in a field twenty-five centimetres below the surface during eight hours; it lived four days after its disinterment (Bardinet, *loc. cit.*). A child was born apparently dead; attempts to revive it proving fruitless, it was laid on a bed for several hours, and then put in a coffin. Twenty-three hours after birth it was seen, and, to make matters certain, the stethoscope was applied to the cardiac region, and heart sounds were heard. All attempts to revive it, however, failed, and it died (Maschka, *loc. cit.*). Such a case as the last suggests the greatest caution, the more so as the body has been opened under similar conditions, and the heart found beating.

The *prognosis* in cases of stillbirth is of the utmost importance as a guide to treatment, but it is not usually made. To attempt to resuscitate a child which is actually dead is useless; though, if the child is fresh, attempts may even here be made for a short time, for the satisfaction of the parents. The certainty of death in the case of fresh children depends on the certainty of the cessation of the heart's action during a considerable period, for the heart may cease to beat for a time, apart from absolute death. In such a case, when the heart has ceased to beat for a considerable time (say ten to fifteen minutes), it might be well cautiously to insert a needle into its apex, before abandoning all efforts at resuscitation.

The question here arises, How long should attempts at resuscitation be continued on the sole ground of the persistent action of the heart? A child whose neck is broken, and who dies eventually with considerable effusion of blood into its spinal cord and brain, has been yet revived and lived for several hours (*vide supra*, p. 78). The case above quoted, in which the heart was found beating twenty-three hours after supposed death, should also be borne in mind. In children who eventually die, the heart revives for a time under artificial respiration, and can be kept beating for some (*e.g.* two) hours, though the child never draws a breath. Such a case, however, is practically hopeless. It may be stated that if no attempts at spontaneous breathing occur within an hour, and especially if the heart, in spite of artificial respiration, acts with diminishing strength and frequency, the prognosis becomes hopeless.

The second point in the prognosis depends on the diagnosis of the state of asphyxia. In the livid stage the prognosis is usually favourable. It is in itself the less serious form, and possesses the great advantage of retaining the power of reflex excitability. If the heart is beating fairly, it is often sufficient to lay the child on its face, wipe out its mouth, and rub it along the spine, a

far better way of exciting an inspiration than slapping
the nates. A livid colour in a stillborn child is, therefore,
a good sign. In the pale or flabby stage of asphyxia the
prognosis is far more serious. These are the cases which
require all the skill at our disposal, while the slight cases
will recover under any treatment, and generally better
without any artificial respiration at all ; it is these latter
cases which form the bulk of those recorded in the jour-
nals as " successful treatment of suspended animation by
a new method."

In the cases of pale or flabby asphyxia, the prognosis
depends on our power to raise the child out of its flabby
condition, and render it amenable to reflex irritation.
Until this is done, all rubbing, slapping, bathing, etc.,
are simple waste of time. While the heart beats regularly, we may still hope for recovery up to a reasonable
length of time, as we have observed above.

The state of the pupils furnishes an important element
in prognosis. In profound asphyxia they are widely
dilated, as in death. The re-establishment of the circulation produces no effect on them, but on the re-establishment of respiration they at once contract (Boehm, *loc.
cit.* S. 91).

If the child is breathing, its respiration may be spasmodic, regular, or mixed, periods of spasmodic and
regular respiration alternating. The spasmodic respiration is imperfect respiration, but is gradually replaced
in most cases by rhythmical respiration, and the prognosis is good. If spasmodic breathing replaces regular
breathing, the reverse is, of course, true (*vide supra*,
p. 92, and " Med. Chir. Proceedings," new series, Vol. I.
p. 24).

A few words may here be said on the vexed question of
the treatment of the navel-string. It appears that a
child gains some four to six ounces of blood after birth,

the principal object of which is probably to furnish an additional supply for the newly-established pulmonary circulation. This blood is drawn into its body by the first inspirations and forced into it by the contractions of the umbilical vessels and their ramifications, which begin from the placenta. To deprive a child of seven pounds weight of this amount of blood is the same thing as to deprive an adult, weighing ten stone, of five to seven and a half pounds of blood—a very serious bleeding. In ordinary cases, therefore, it is best to tie the cord late. In cases of asphyxia the same rule holds good, unless manipulations are indicated which require the child to be free from its mother.

It has been recommended, in cases of livid asphyxia, to allow half an ounce to an ounce of blood to escape from the cord. This treatment is probably founded on the full-blooded appearance of the child. The child, however, has *less* blood than it should eventually have, and has no more than a child in the pale stage, the difference being one in the distribution rather than the amount of blood. Remembering that these livid children generally recover, and that the treatment has never been suggested for the pale children, who are the really serious cases, it is probably best, in the meanwhile at least, not to bleed.

The colour of the skin, and especially of the lips, is important, whether tending to lividity or to pallor, and a favourable change often precedes the establishment of respiration.

With regard to artificial respiration, it may be stated that four objects are aimed at, namely:—

(1) Removal of foreign bodies from the air-passages ;

(2) Procuring the patency of the air-passages;

(3) Excitation of the circulation;

(4) Ventilation of the lungs.

(1) With regard to the first, the mouth, and, as far as possible, the air-passages, should be cleared before inspiratory movements are excited, in order to prevent the liquor amnii, meconium, vernix caseosa, and other matters from being inhaled into the lungs. The best method to pursue is the following : Lay the child on its back, with the head hanging over the edge of a table, a little lower than the rest of its body. Wipe out the mouth with a soft handkerchief. Press the thorax gently with one hand, stroking the trachea upward with the other, and retain the finger at the top of the trachea until the next manœuvre is complete. The mucus will gravitate towards the posterior nares. Put a handkerchief over the child's mouth, blow gently, and the mucus will be blown out of the nostrils, but not into the operator's face (*vide supra,* p. 90).

If there is great accumulation of mucus in the air-passages, a No. 9 gum-elastic male catheter should be introduced into the trachea, so that the point is three and a half inches from the lips. This length will secure its passing through the glottis, but not as far as the bifurcation of the trachea. Press the thorax gently with one hand to prevent the entrance of air, and blow through the catheter. The opening being low down in the trachea, the air and mucus with it, being unable to pass into the lungs on account of its compression by the hand, will rush up through the glottis, and the mucus will be blown into the pharynx. This can be repeated as often as necessary, the general tendency of fluids in the air-tubes being to ascend during respiration, whether natural or artificial, towards the mouth. This manœuvre is more efficient and far pleasanter than the suction usually recommended ; it has answered well in practice (*vide supra,* p. 90).

(2) With regard to securing the patency of the air-

passages; this is often the greatest difficulty in artificial respiration, particularly where the child is deeply asphyxiated, and the divaricator muscles of the glottis seem to share the general flaccidity of the muscles of the whole body (*vide supra*, p. 89). All methods of artificial respiration by manipulation, except that of Schultze, often seem to be useless, on account of this condition; the method of Schultze, however, seems by some means to render the air-passages patent in many cases where they have remained closed during other manipulations (*vide supra*, p. 89). To pull the tongue forward with forceps is useless, as it does not affect the structures at the base of the tongue, and does not raise the epiglottis (Howard, *loc. cit.*). The manœuvre of tilting up the chin with the mouth closed, however useful in adults, permitting respiration solely through the nose, as it does, cannot be of the same use in stillborn children, whose pharynx and air-passages are filled with mucus, and in whom the removal of mucus is one of the great desiderata (*vide supra*, p. 88). Hanging the head backwards over the edge of a table does not provide for the patency of the upper air-passages (*vide supra*, p. 89). If the difficulty is persistently experienced, a catheter can be introduced into the trachea, and secured at the proper length of three and a half inches, so that it may not slip too far in, or slip out during manipulations. While in the trachea it serves for various purposes, among which is the removal of mucus as described above. During artificial respiration fluid will often pour from the catheter, especially if it is held over the edge of the table like a siphon (*vide supra*, p. 91).

(3) The third object, excitation of the circulation, is generally dependent on the fourth, namely, ventilation of the lungs. But pressure over the præcordia has a direct effect in raising the blood-pressure and exciting the

action of the heart (Boehm, *loc. cit.* S. 72 *et seqq.*). This is probably the chief reason for any success obtained by the methods of Howard and Marshall Hall, which are powerless to introduce air into the lungs of stillborn children (*vide supra*, p. 49). It has been said that it is undesirable to excite the circulation in deeply asphyxiated children, since the blood which is supplied to the body is necessarily impure. This objection, however, is theoretical, the truth being that the revival of the circulation gives the best prognosis for the establishment of respiration.

(4) The ventilation of the lungs is secured by various methods of manipulation, of which, for stillborn children, only two are trustworthy, namely, the method of Schultze, and that of Silvester, with its modifications by Pacini and Bain (*vide supra*, p. 49). It may be said at the outset that the method of Silvester (and its modifications), though the best for general purposes, by no means imitates the natural breathing of an infant, which is diaphragmatic (*vide supra*, p. 47). In the method of Schultze, on the other hand, the diaphragm descends, though but slightly (*vide supra*, p. 46). The chief effect of both methods is to raise the upper part of the thorax. There remains the method of mouth-to-mouth inflation, the object of which is to fill the lungs once for all.

These methods will now be described.

A. The method of Schultze (*loc. cit.* S. 162, and *supra*, p. 8). The navel-string being tied, the child is seized with both hands by the shoulders in such a way that both thumbs lie on the anterior wall of the thorax, both index fingers extend from behind the shoulders into the axillæ, and the other three fingers of both hands lie obliquely along the posterior wall of the thorax. The head is prevented from falling by the support of the ulnar sides of the two hands. The operator stands with somewhat separated legs, and bends slightly forward, holding the child, as above described, at arm's length, hanging perpendicularly. (*First position, inspiratory.*)

Without pausing, he swings the child upwards from this hanging position at arm's length. When the operator's arms have gone slightly beyond the horizontal, they hold the child so delicately that it is not violently hurled over, but sinks slowly forwards and forcibly compresses the abdomen by the weight of its pelvic end. (*First movement, expiratory.*) At this moment the whole weight of the child rests on the operator's thumbs lying on the thorax. (*Second position, expiratory.*) Any compression of the thorax by the hands of the operator must be carefully avoided. The body of the child rests during the first position with the floor of the axillæ on the index fingers of the operator exclusively, and no compression should be exercised on the thorax in spite of the support of the hands to the head, nor should the thumbs compress the thorax in front. When the child is swung upwards the spinal column should not bend in the thoracic, but only in the lumbar region, and the thumbs should not at this time strongly press the thorax, but should only support the body as it sinks slowly forward. The raising of the body as far as the horizontal should be effected by a powerful swing of the arms (of the operator) from the shoulders, but from that point the arms should be raised more and more slowly, and, by means of a delicately adjusted movement of the elbow-joints and scapulæ on the thorax, the pelvic end of the child should fall gradually over. By this gradual falling of the child's pelvis over the belly, considerable pressure of the thoracic viscera is effected, both against the diaphragm and the whole thoracic wall. At this point the inspired fluids often pour copiously from the respiratory openings. After the child has slowly but completely sunk over, the operator again lowers his arms between his separated legs. The child's body is thereby extended with some impetus ; the thorax, released from all pressure (the operator's thumbs lying now quite loosely on the anterior walls of the chest), expands by means of its elasticity, but the weight of the body hanging, as it does, on the index fingers of the operator by the upper limbs, and thus fixing the sternal ends of the ribs, is brought into use for the elevation of the ribs with considerable impetus ; moreover, the diaphragm descends by virtue of the impulse which is communicated to the abdominal contents.

(*Second movement, inspiratory.*)   After a pause of a few seconds, in the first inspiratory position, the child is again swung upwards into the previous position (*first movement, second position, expiratory*), and while it sinks slowly forwards it brings its whole weight to bear on the thumbs, which rest on the anterior thoracic wall, and mechanical expiration again ensues.   At this point any inspired fluids always pour copiously from the mouth and nose, and generally meconium from the anus.   The proceeding is repeated eight or ten times a minute, but more slowly when the inspired fluids flow from the mouth and nose.

It is most important that at the end of the respiratory movement the weight of the child's body should be entirely thrown on the index fingers placed in the axillæ, and none of it supported by the rest of the hand (*vide supra*, p. 44).

Schultze's method has the disadvantage of being more sudden and violent than some of the other methods (*vide supra*, p. 44), but this can be largely controlled by the operator.   On the other hand, it possesses two great advantages : the first is, that in the expiratory position the child is inverted, and gravity assists in removing mucus and other bodies from the air-passages ; the other is, that for some reason it does sometimes actually procure patency of the air-passages where other methods fail (*vide supra*, p. 89).

The method of Schultze has recently been assailed on the ground that it fails to introduce air into the lungs. It would be very difficult to devise an experiment which would prove or disprove this.

In the experiments on stillborn children referred to above, tracheotomy had been performed and a cannula tied into the trachea, which really removes the chief impediment to the introduction of air in actual practice, such impediment concerning almost exclusively the upper air-passages. As regards the results of autopsies, it must be remembered that the presence of a certain amount of air in the lungs of a stillborn child would not prove its introduction by artificial respiration, since the lungs of children dying *intra partum* are not necessarily airless (Schwartz, *loc. cit.*; Schultze, *loc. cit.* S. 132; Hecker, *loc. cit.*), but complete, or even considerable expansion of the lungs of a stillborn child after artificial respiration would certainly prove that air had been introduced by such artificial respiration. It is doubtless true that in some cases all methods will fail, and that all manipulative methods often fail in the case of premature or undersized infants, unless, indeed, the patency of the upper air-passages is secured. It is not enough to prove that in some cases a method succeeds—what is desired is to know the best methods and what to expect of them. The result of considerable experience of all methods of artificial respiration in practice has been to assign to Schultze's method one of the highest places as a means of securing the ventilation of the lungs (*vide supra*, p. 38 and p. 50), and the highest place among manipulative methods as a means of securing patency of the upper air-passages (*vide supra*, p. 89).

B. The method of Silvester (*loc. cit.* and *supra*, p. 6), recommended by its author for children and adults:

(1) To adjust the patient's position, place the patient on his back, with the shoulders raised and supported on a folded article of dress; (2) to maintain a free entrance of air into the windpipe, draw the tongue forward; (3) to imitate the movements of respiration raise the patient's arms upward by the sides of his head, and then extend them gently and steadily upward and forward for a few moments, next turn down the patient's arms and press them gently and firmly for a few moments against the sides of the chest; (4) the feet are to be secured, and the arms are to be stretched steadily upward for two seconds. The arms should, if possible, be everted, which gives a greater inspiratory power, by rendering more tense the tendons of the pectoralis major muscles, and they should be seized above the elbows (*vide supra*, p. 43).

The modifications by Pacini and Bain are as follows. That by Pacini is thus described (*loc. cit.* and *supra*, p. 7) : The feet of the patient being fixed, the operator stands with the head against his own abdomen, and then with his hands takes a firm hold of the upper part of the arms, applying the forefingers behind and close to the armpits, while the thumb is in front of the head of the humerus. Holding the shoulders thus, he pulls them towards him, and then lifts them in a perpendicular direction.

Bain's modification is the following (*loc. cit.* and *supra*, p. 7). First method : The fingers are placed over the front of the axillæ, the thumbs over the ends of the clavicles; the operator then draws the shoulders upward and then relaxes his traction. Second method : The shoulders are raised by taking hold of the hands and raising the body about a foot off the table, the position of the arms being at an angle of about forty-five degrees beyond the head. These modifications present no practical advantages over the original method of Silvester for children ; in adults they seem to increase slightly the amount of air inspired.

The advantages of Silvester's method are its simplicity and the comparatively large amount of air which it is capable of drawing into the lungs. Its disadvantages are that the supine position of the patient impedes the escape of mucus and other matters from the mouth, and also the fact that it frequently happens that the collapsed state of the upper air-passages prevents the introduction of any air into the lungs (*vide supra*, p. 89).

he methods of Marshall Hall and Howard (*vide supra*, pp. 40, 41, 49) are incapable of introducing any air into the lungs of a stillborn child, because the chest has no resiliency, and any effect they may produce in resuscitating stillborn children must be produced by pressure on the heart, which has the power of stimulating its action and raising the blood-pressure (Boehm, *loc. cit.* p. 91).

We have now to say a few words on direct inflation of the lungs. Various curved cannulas have been invented for this purpose, especially in France, but we cannot recommend their addition to the already large armamentarium of the accoucheur ; and, if used at all, they must be constantly carried, for their need arises unexpectedly. The best method is that by the mouth. The operation may be rendered clean by the simple method of laying a towe over the child's mouth and breathing through it. The disadvantages of direct mouth-to-mouth inflation are alleged to be the following : (1) Danger of rupturing the lungs ; (2) danger of tubercular infection from the operator ; (3) danger of inflating the stomach and so preventing inspiration by impeding the descent of the diaphragm.

The first can be avoided by gentleness in inflating, and also by leaving the nose free. To close the nostrils, as is often recommended, is useless and dangerous—the nostrils are a valuable safety-valve (*vide supra*, p. 91).

L

The second is founded on an observation by Reich (*loc. cit.*), relating to a consumptive midwife who practised inflation. In thirteen months twelve children delivered by her were attacked with symptoms of bronchial catarrh with fever, and died of tubercular meningitis. The children had no phthisical family history. During the same period the other midwife of the place had no cases of tubercular meningitis. In the nine years previous to the practice of the phthisical midwife there had been only two cases, and in the year after her death only one. In view of recent researches on tubercle, the above has at least to be remembered.

The third is imaginary, so far as artificial respiration is concerned; air in the stomach in no way diminishes the amount of air which can be introduced into the lungs by the artificial methods, and it is easily pressed up from the stomach (*vide supra*, p. 86).

It has been suggested that the entrance of air into the stomach can be prevented by pressing the cricoid cartilage against the bodies of the vertebræ, and also by bending the head well back. Neither of these plans produces the desired effect in children, though the former succeeds in adults (*vide supra*, p. 87).

It remains for us to say a few words on the action of heat. In this matter it is necessary to keep apart the two questions of (*a*) recovery to respiration, and (*b*) avoiding chills. It has been proved that, within limits, the lower the temperature the longer can an animal survive without breathing, and the higher the temperature the more quickly it dies (Edwards, *loc. cit.*). To keep a child in a hot bath until respiration is established is, therefore, a wrong practice. The hot bath can, however, be advantageously used in alternation with the cold bath, but merely as a means of increasing the effect of the cold bath, and in the pale (flabby) stage of asphyxia this also is useless. Under such circumstances it is best to wrap the child in a warm flannel, and not to waste time on baths, but to proceed at once to the establishment of respiration. The application of warm flannels to the head, which is a valuable nervous stimulant in the case of adults, may be advantageously tried in the case of infants, so soon as any signs of reflex action appear, or even possibly earlier.

To sum up the treatment, we may say : Never hurry, it is not a question of seconds, and success depends upon a fine exercise of the judgment. Make a good diagnosis, first, as to life or death; secondly, as to the stage of asphyxia (if life is not extinct). If the child is macerated, it is obviously dead and past hope. If the heart beats ever so slowly and feebly, it is not dead. If the heart is not beating, death is not certain, unless it can be proved to be inactive for some time. If the child is livid and not flabby, it will probably come round; wipe out its mouth and pharynx, and rub it with a soft cloth down the spine, press gently on the cardiac region. If this produces no effect, or if it be in the pale stage, inflate the lungs by the mouth, and then by Silvester's method. If air enters the lungs, well and good; if not, try Schultze's method, or insert a catheter, as described above. On the first sign of muscular action, plunge the child into cold water, or into alternate hot and cold baths. Vary the treatment between occasional inflation of the lungs, artificial respiration, pressure over the cardiac region, baths, irritation down the spine, according to the judgment; remembering what may be expected of each method, and that no one will suffice for all cases. Watch for signs of resuscitation, namely, improvement in the colour, in movements, in cardiac pulsations, as described above. Never be content until the child breathes regularly and appears to be continually improving.

L 2

List of Works quoted.

Bain, "Med. Times and Gazette," Dec. 19, 1868, p. 708.

Bardinet, "Bulletin de l'Acad. Imp.," tome xxx. p. 1052, No. 21, 15 août, 1865 (quoted by Schultze).

Boehm, "Arch. für exp. Path. und Pharm.," Band viii. S. 72, *seqq.*

Bohn, "De officio Medici Forensis," S. 262 (quoted by Schultze).

Cazeaux, "Gaz. Méd." 1850, No. 17, p. 317, and "Traité des Accouchements," Bruxelles, 1845, p. 34.

Champneys (F. H.), "Med. Chir. Trans.," vol. lxiv. 1881, pp. 41–86.

Champneys (F. H.), Ibidem, vol. lxvii. 1884, pp. 106–112.

Edwards (W. F.), "Annales de Chimie," 1818, tome viii. série 2, p. 225.

Hecker, "Virch. Archiv," 1859, xvi. S. 534.

Howard, "Lancet," May 22, 1880, p. 797.

Litzmann, "Die Geburt bei engen Becken," Leipzig, 1884.

Maschka, "Das Leben der Neugeborenen ohne Athmen," Prager Vierteljahrschrift, 43, S. 1, 1854.

Pacini, "Di un nuovo Metodo di practicare la Respirazione artificiale," Firenze, 1867.

Reich, "Berl. Klin. Woch.," 1878, No. 37.

Schwartz, "Die vorzeitigen Athembewegungen," Leipzig, 1858, S. 136 (Cases 15 and 30).

Schultze, "Der Scheintod Neugeborener," Jena, 1871.

Silvester, "The true physiological Method of restoring Persons apparently Drowned or Dead, and of resuscitating Stillborn Children," 1858, p. 17.

INDEX.

—•—

Air, presence of, in abdominal viscera (stomach) as an impediment to artificial respiration, 85, 92, 146

,, presence of, in lungs of children dying during birth, 143

Anterior surfaces of lungs often better expanded than the posterior, 61

Apices of lungs often the least expanded part, 61

"Apoplectic" state of stillbirth, 132

Apparatus, description of, employed in testing the amount of ventilation of the lungs secured by each method, 10

Arms, eversion of, during Silvester's method important, 43, 144

Asphyxia, stages of ("livid" and "pale"), 2, 132, 135

Atelectasies, small, Schultze's claim to have detected and removed them, 45

Bain, his method described, 7, 144

,, method of, practically identical with those of Pacini and Silvester, 43, 144, 149

,, second method of, useless, 43, 49

Baths, hot and cold, in treatment of stillbirth, 146

Behm (Carl), 3

Behm (Carl), discussion and criticism of his experiments, 47

Bleeding from umbilical cord as a remedy in asphyxia, 136

Blood, effused, inspiration of, into thorax by artificial respiration, 78

Boxall, Dr. Robert, case of cervical emphysema, 128

Bronchi and trachea, impossible to burst when healthy, 108, 109, 111, 112, 113, 114, 123, 126

Catheter, use of, in securing patency of upper air-passages, 90, 138, 139

Cervical emphysema, expiratory, 94

Circulation, excitation of in asphyxia, 139

Collapse, position of, in chest of newborn child, 49

Colour of skin as bearing on diagnosis and prognosis, 136, 137

Conclusions on amount of ventilation secured by various methods, 49, 140

Cord, umbilical, treatment of, at birth, 136

Cricoid, pressure on, does not occlude œsophagus without also occluding trachea, 87, 92, 146

Desiderata of methods of artificial respiration, 3, 137

Diaphragm, descent of, by Schultze's method, 26, 27, 28, 46, 50, 140

Emphysema (mediastinal) and pneumothorax, 18
,, connected with Schultze's method, 45
,, of anterior mediastinum in artificial respiration after tracheotomy, 61
,, in connection with tracheotomy, 68, 81
,, after tracheotomy, frequency of, 82
,, clinical observation of (Dr. Douglas Powell), 83
,, expiratory cervical, 94
,, of neck after labour, frequency of, 95, 126
,, ,, aetiology of, 95, 125
,, ,, clinical course of, 96
,, ,, cases of, 96, 128
,, ,, pathology of, 98
,, ,, experiments illustrative of, 100
,, ,, experiments considered, 119
,, ,, route selected by air in, 119
,, ,, source of air in, 119
,, ,, production of, explained, 120, 125
,, ,, conclusions concerning, 126, 127
,, lower part of trunk entirely distinct from that of neck, 126

Error, sources of, in experiments, eliminated, 4

Expanded patches in lungs not bounded by fissures, 61

Expansibility of various parts of the lungs under artificial respiration, 51 et seqq.

Experiments to test the amount of ventilation secured by each method, 11

Expiratory cervical emphysema, 94

Expiratory force of each method important, 50

Faradisation of phrenic nerves, 49

Foreign bodies, removal of, from air-passages, 90, 138

Glottis, paresis of divaricators of, as an impediment to respiration, 89, 139

Hall, Marshall, his method described, 5
,, method of, useless, 39, 41, 145
,, reason of uselessness of method of, for stillborn children, 49, 145

Head, hanging of, backwards, does not secure patency of upper air-passages, 139

Heat, action of, in asphyxia, 146

Howard, his method described, 6
,, method of, useless, 39, 40, 41, 49, 145
,, reason of uselessness of method of, 40, 145
,, method of effecting patency of upper air-passages not suitable for infants, 88, 139

Inflation, direct, of lungs, 145
,, dangers of, 145-6

Joerg, analysis of his observations, 65

Legs, fixation of, important in Silvester's method and its modifications, 49, 144

Life, signs of returning, after asphyxia, 91, 136

Lungs, anterior surfaces of, often better expanded than the posterior, 61

,, apices of, often the least expanded part, 61

,, right, generally better expanded than left, 61

,, unexpanded strip running vertically opposite costal angles, 61

,, least expanded parts of, 62

,, less expansion of posterior part not due to hypostatic congestion, 62

,, most and least expanded parts of, underlie most and least mobile parts of thorax, 63, 67

,, condition of, in children dying soon after birth, 66

,, condition of, in stillborn children after artificial respiration (conclusions), 66

,, of fœtus permeable by fluids when unpermeable by air, 104, 124, 126

,, order of expansion of parts when removed from chest, 107, 109

,, comparative tenacity of different parts of, 108, 109, 110, 111, 115, 117, 119

,, pressure necessary to produce rupture of, 110, 112, 113, 115, 117, 119, 122

,, weak spots in, 119, 125

,, behaviour of, when thorax was opened, 110, 111, 112, 113

,, root of, weakness of to distension, 120, 126

Marshall Hall (see Hall, Marshall)

Material used in this investigation, 4

Mediastinal emphysema in artificial respiration with tracheotomy, 61

,, and pneumothorax in connection with tracheotomy, 68, 81

,, and pneumothorax after tracheotomy, frequency of, 82

,, clinical observation of (Dr. Douglas Powell), 83

Methods of artificial respiration employed, classification of, 5

,, description of, 5–10

,, list of, 4

Minor points in artificial respiration, 85, 92

Mucus, removal of, best method, 90, 138

"Necessity of breathing" at birth, 133

Neck, root of, precautions to be observed in surgery of, 80

Nostrils, safety-vale action of, during mouth-to-mouth inflation, 91, 145

,, should not be held during mouth-to-mouth inflation, 91, 145

Œsophagus is not satisfactorily occluded by pressure on cricoid, nor by bending head back, 87, 92, 146

Opisthotonos produces expiration, 50

Pacini, his method described, 7, 144

,, method of, practically identical with that of Bain and Silvester, 43, 49, 144

,, latter half of method of, useless, 43

,, uselessness of second half of his method, 49

"Pale" stage of asphyxia the more serious, 2, 3, 136

Patches of expansion in lungs not bounded by fissures, 61

Patency of upper air-passages, best means of securing, 138

,, not secured by hanging head backwards, 88, 92

,, nor by pulling tongue forwards, 139

,, best secured in method of Schultze, 89, 139

Phrenic nerve, descent of air from tracheotomy wound along course of, 70, 73

Pleura, tenacity of pulmonary, 108, 109, 110, 112, 115, 119, 122

Pneumothorax and emphysema connected with Schultze's method, 45

,, in artificial respiration, 61

,, and mediastinal emphysema in connection with tracheotomy, 68, 81

,, and mediastinal emphysema after tracheotomy, frequency of, 82

,, clinical observation of (Dr. Douglas Powell), 83

,, not connected with expiratory emphysema, 126

Powell, Dr. Douglas, on mediastinal emphysema and pneumothorax after tracheotomy, 83

Præcordia, pressure on, as a stimulant to circulation, 139

Prognosis in cases of stillbirth, 134

Pupils, state of, in stillbirth, 136

Reflex action abolished in "pale" stage of asphyxia, 2, 132

Respiration, spasmodic, mixed and regular in asphyxia, 136

Respiration, artificial, objects of, 137

Right lung generally better expanded than left, 61

Rotation (outward) of arms during Silvester's method, 19, 21, 144 et passim

Schmitt, analysis of his observations, 64-5

Schrœder, his method described, 8

,, method of, useless, 39, 40, 50

,, rationale of method of, 40

,, method of, reason of uselessness of, 40

Schücking, his method described, 7

,, method of, practically identical with that of Silvester, 42

,, method of, no improvement on Silvester's, 50

Schüller, his method described, 8

,, method of, useless, 41

,, reason of uselessness of method of, 41

,, method of, useless, and not free from risk, 50

Schultze's method described, 8-10, 140

,, importance of minutiæ in, 28, 44, 142

,, rationale of, 44

,, emphysema and pneumothorax connected with, 45

,, efficient, 50, 143

,, descent of diaphragm in, 50, 140

,, principal action on thoracic walls, 50

,, importance of details in, 50, 142

,, violence of, 15, 44, 50, 142

,, emphysema of anterior mediastinum and pneumothorax most apt to occur during, 74, 80

,, best secures patency of upper air-passages, 89, 139

"Schultze-Silvester" method defined, 12

,, description of, 46

Silvester, his method described, 6, 143

,, general consideration of method of, 42, 145

,, modifications by Pacini, Bain, and Schücking give no advantage over Silvester's original method, 42–44, 144

,, method of, unlike natural respiration of a child, 47, 140

,, produces most effect in ventilating lungs, 49, 145

,, importance of details in, 49, 144

Stillbirth defined, 132

Stillborn children who have never breathed, position of equlibrium of chest of, 40

Spirophore of Woillez, 49

Strip, unexpanded, of lungs running vertically opposite costal angles, 61, 132

Survival, instances of, in stillbirth, 134

"Syncope," state of, in stillbirth, 132

Table showing expansibility of various parts of lungs, 54 et seqq.

Thorax, difference in shape and pressural relations of, at, and soon after birth, 47, 48

,, collapsed state of, in newborn child, 49, 145

Tongue, pulling orward of, useless as a means of securing patency of the upper air-passages, 88, 139

Trachea and bronchi, impossible to burst when healthy, 108, 109, 111, 112, 113, 114, 123, 126

Tracheotomy, mediastinal emphysema and pneumothorax in connection with, 68, 81

,, post-mortem records of, in St. Bartholomew's Hospital, 74

,, at the Hospital for Sick Children, 75, 82

, precautions to be observed during, 79

,, dangerous period during, 79

,. frequency of mediastinal emphysema and pneumothorax after, 82

,, clinical observation of (Dr. Douglas Powell), 83

Treatment, summary of, 147

Tubercular infection by inflation of lungs, 146

Uselessness of methods of Schrœder, Howard, Marshall Hall, and Schüller, 39, 41, 145

Ventilation of lungs, amount of, secured by different methods of artificial respiration, 1

,, best practical methods, 140

,, summary of experiments on, 38

Wilks (and Moxon) on pneumothorax after tracheotomy, 76

Woillez, spirophore of, 49

H. K. LEWIS, Printer, 136, Gower Street, London, W.C.

www.ingramcontent.com/pod-product-compliance
Lightning Source LLC
Chambersburg PA
CBHW021809190326
41518CB00007B/521